Python Essentials for AWS Cloud Developers

Run and deploy cloud-based Python applications using AWS

Serkan Sakinmaz

BIRMINGHAM—MUMBAI

Python Essentials for AWS Cloud Developers

Group Product Manager: Mohd. Riyan Khan
Publishing Product Manager: Suwarna Rajput
Senior Editor: Runcil Rebello
Technical Editor: Rajat Sharma
Copy Editor: Safis Editing
Project Coordinator: Ashwin Kharwa
Proofreader: Safis Editing
Indexer: Tejal Daruwale Soni
Production Designer: Arunkumar Govinda Bhat
Marketing Coordinator: Agnes D'souza

First published: May 2023

Production reference: 1260423

Published by Packt Publishing Ltd.
Livery Place
35 Livery Street
Birmingham
B3 2PB, UK.

ISBN 978-1-80461-006-0

www.packtpub.com

To my mother, Reyhan, and my father, Sami, for always supporting and loving me. To my sons, Batu and Arman, for recharging my energy. To my wife, Yonca, for giving me support and love.

– Serkan Sakinmaz

Contributors

About the author

Serkan Sakinmaz is a data architect and engineer who lives in Germany. He currently gives consultancy on the data and cloud area to key companies in Europe. He has also given big data training to students as well as professionals who want to learn about cloud technologies. He has more than 15 years' experience in programming and more than 8 years' experience in the cloud area. He likes to share what he knows and his experiences in the sector, and he gives seminars, writes blogs, and defines the future of architecture for key companies.

When he is not working, he mostly spends his time with his family, goes running, or plays table tennis and football.

About the reviewers

Harish LM has more than 5 years of experience in the IT and services industry. He specializes in Python, NLP, and the AWS Cloud.

Mikayel Ghazaryan is a technology expert with extensive experience in web and cloud engineering and software design. He specializes in designing and implementing scalable, secure, and cost-effective solutions for businesses. He currently works as an AWS Cloud architect at Nordcloud, an IBM company since December 2021. His responsibilities include performing a Well-Architected Review, and standardization of EC2 Linux machine deployments with Terraform, Systems Manager, Step Functions, and Lambdas. Before Nordcloud, he performed the migration of on-premise applications to AWS, designed and implemented data pipelines on AWS, and attributed ML-based quality scores to millions of images. He holds an AWS Solutions Architect Professional certification.

Table of Contents

4

Running Python Applications on EC2   41

5

Running Python Applications with PyCharm   65

6

Deploying Python Applications on Elastic Beanstalk   75

Part 3: Useful AWS Services to Implement Python

Preface

Cloud computing is one of the most popular approaches to implementing your applications, with huge advantages. There are multiple cloud providers, such as AWS, GCP, and Azure. AWS is one of the most used cloud providers, and many companies are moving there. Cloud usage is significantly growing and cloud knowledge is expected from developers.

Most of the applications are moving to the cloud. AWS has different services to implement Python applications, hence the configuration and selecting the right service is a challenge for those who don't have an AWS background. By buying this book, you are on the right path and stepping into how to implement cool Python applications using AWS services.

Who this book is for

This book is implemented for cloud developers, software developers, and IT specialists who intend to develop Python applications on AWS as well as learn about the concepts of appropriate AWS services for implementing the Python applications. You should have Python programming experience to implement the applications on AWS.

What this book covers

Chapter 1, Using Python on AWS. This chapter will teach you how to install and use the Python IDE and also understand the advantages of AWS Cloud.

Chapter 2, Creating an AWS Account. To start with cloud computing, AWS requires an account to implement Python programming. In this chapter, you will learn how to create an AWS account.

Chapter 3, Cloud Computing with Lambda. Lambda is a very effective way to implement Python functions. The chapter will help you to get into the Lambda service and will show how to implement a code.

Chapter 4, Running Python Applications on EC2. EC2 is one of the key services that you can provision on the cloud. The chapter will help you to get into the EC2 service and will show how to provision a server and deploy the Python application afterward.

Chapter 5, Running Python Applications with PyCharm. Debugging Python applications is important for testing the application. The chapter will help you to debug Python applications locally in an easy way.

Chapter 6, Deploying Python Applications on Elastic Beanstalk. Elastic Beanstalk is a useful service that allows the deployment of applications. The chapter will help you to get into the Elastic Beanstalk service and will show how to create a service and deploy the Python application afterward.

Chapter 7, Monitoring Applications via CloudWatch. CloudWatch allows you to monitor your application in AWS. The chapter will help you to get into the CloudWatch service and will show how to monitor the Python application.

Chapter 8, Database Operations with RDS. RDS is used to create a database in AWS. The chapter will help you to get into the RDS service and will show how to create a database and make SQL operations via Python applications.

Chapter 9, Creating an API in AWS. An API is an important interface for an application. The chapter will help you create an API in AWS and publish the API to access the Python application.

Chapter 10, Using Python with NoSQL (DynamoDB). NoSQL is useful to store unstructured and semi-structured data. The chapter will help you to create a NoSQL database and make SQL operations on DynamoDB.

Chapter 11, Using Python with Glue. Glue is a serverless data integration service in AWS. The chapter will help you to embed Python applications into the Glue service.

Chapter 12, Reference Project on AWS. Implementing a sample project is the best way to learn about application programming. The chapter will help you to implement sample AWS projects with best practices.

To get the most out of this book

You will need to have an understanding of the basics of the Python programming language to implement applications on AWS.

Software/hardware covered in the book	Operating system requirements
Python	Windows, macOS, or Linux
Amazon Web Services (AWS)	

Download the example code files

You can download the example code files for this book from GitHub at https://github.com/PacktPublishing/Python-Essentials-for-AWS-Cloud-Developers. If there's an update to the code, it will be updated in the GitHub repository.

We also have other code bundles from our rich catalog of books and videos available at https://github.com/PacktPublishing/. Check them out!

Download the color images

We also provide a PDF file that has color images of the screenshots and diagrams used in this book. You can download it here: `https://packt.link/hWfW6`

Conventions used

There are a number of text conventions used throughout this book.

`Code in text`: Indicates code words in text, database table names, folder names, filenames, file extensions, pathnames, dummy URLs, user input, and Twitter handles. Here is an example: "Execute `python --version` from the command line."

A block of code is set as follows:

```
from flask import Flask
app = Flask(__name__)

@app.route('/')
```

When we wish to draw your attention to a particular part of a code block, the relevant lines or items are set in bold:

```
from flask import Flask
app = Flask(__name__)

@app.route('/')
```

Any command-line input or output is written as follows:

```
wget https://raw.githubusercontent.com/PacktPublishing/Python-
Essentials-for-AWS-Cloud-Developers/main/fileprocessor.py
```

Bold: Indicates a new term, an important word, or words that you see onscreen. For instance, words in menus or dialog boxes appear in **bold**. Here is an example: "Click **Instances** on the left side, and then click **Launch Instances**."

> **Tips or important notes**
> Appear like this.

Get in touch

Feedback from our readers is always welcome.

General feedback: If you have questions about any aspect of this book, email us at customercare@packtpub.com and mention the book title in the subject of your message.

Errata: Although we have taken every care to ensure the accuracy of our content, mistakes do happen. If you have found a mistake in this book, we would be grateful if you would report this to us. Please visit www.packtpub.com/support/errata and fill in the form.

Piracy: If you come across any illegal copies of our works in any form on the internet, we would be grateful if you would provide us with the location address or website name. Please contact us at copyright@packt.com with a link to the material.

If you are interested in becoming an author: If there is a topic that you have expertise in and you are interested in either writing or contributing to a book, please visit authors.packtpub.com.

Share Your Thoughts

Once you've read *Python Essentials for AWS Cloud Developers*, we'd love to hear your thoughts! Scan the QR code below to go straight to the Amazon review page for this book and share your feedback.

https://packt.link/r/1804610062

Your review is important to us and the tech community and will help us make sure we're delivering excellent quality content.

Download a free PDF copy of this book

Thanks for purchasing this book!

Do you like to read on the go but are unable to carry your print books everywhere? Is your eBook purchase not compatible with the device of your choice?

Don't worry, now with every Packt book you get a DRM-free PDF version of that book at no cost.

Read anywhere, any place, on any device. Search, copy, and paste code from your favorite technical books directly into your application.

The perks don't stop there, you can get exclusive access to discounts, newsletters, and great free content in your inbox daily

Follow these simple steps to get the benefits:

1. Scan the QR code or visit the link below

https://packt.link/free-ebook/9781804610060

2. Submit your proof of purchase
3. That's it! We'll send your free PDF and other benefits to your email directly

Part 1:
Python Installation
and the Cloud

In this part, you will learn to install and use the Python IDE and understand the cloud basics. In order to get into cloud computing via Python programming in AWS, we will also open an AWS account.

This part has the following chapters:

- *Chapter 1, Using Python on AWS*
- *Chapter 2, Creating an AWS Account*

1
Using Python on AWS

In this chapter, we will give a brief introduction to the cloud. We will then explain how to set up Python and how to run your first application within the command line as well as via an **integrated development environment (IDE)**. We're going to cover the following main topics:

- What is the cloud?
- Understanding the advantages of the cloud
- Installing Python
- Installing PyCharm
- Creating a new project

Cloud computing is one of the most popular approaches to implementing your applications, and it has huge advantages. There are multiple cloud providers, such as **Amazon Web Services (AWS)**, **Google Cloud Platform (GCP)**, and Azure. AWS is one of the most widely used cloud providers, and many companies are moving there. Cloud usage is significantly growing, and developers are expected to have a good understanding of the cloud. By buying this book, you are on the right path and stepping into how to implement cool Python applications using AWS.

Most companies are moving to the cloud because of the significant advantages. It is important to know why and how these services are being used.

What is the cloud?

The cloud is a popular way of using your IT infrastructure and services over IT providers that manage machines, networks, and applications. Basically, you don't need any on-premises infrastructure, and cloud providers have their data centers to serve the required services over the internet. For example, if you need a server, you don't need to buy a machine and don't need to set up its network and power. Cloud providers serve these resources for you, and you can use them over the internet.

Understanding the advantages of the cloud

The following aspects explain why companies are moving to the cloud to have a better infrastructure:

- **Good disaster recovery plan**: Cloud providers have multiple data centers in different regions. If an issue happens in one region, the system can be recovered in another region.

- **Better scalability and stability**: In AWS, you have different services to upscale and downscale your application. All you need to do is to configure scaling options based on usage.

- **Quicker time to production**: AWS has more than 100 services, and these services come with huge capabilities. When you have any application for production, you don't need to start from the beginning, such as provisioning the server or preparing the infrastructure.

- **Pay-as-you-go model to reduce the cost**: You don't need to sign a contract that promises payment; you can also use the service for just one day and then shut it down.

- **Monitoring and logging advantages**: The biggest cloud providers have monitoring and logging services; you can integrate these services into your application.

- **Reduces DevOps effort**: AWS comes with lots of advantages for DevOps. For example, you can provision servers quickly and deploy and monitor your service with simple configurations.

- **Multiple security services to keep data safe**: There are different services to keep your services and data safe.

The cloud comes with lots of advantages. There are also some important considerations when using cloud services:

- **Security**: Securing your services is important, and AWS provides different services to protect your data, such as firewall configurations. You have to evaluate security requirements while using AWS services.

- **Cost management**: You can easily create and scale your services, which is a very big advantage. The point to note is that while you create these services, it comes with a cost, which can cause surprises if you don't consider the costs for specific services. Check the cost of services while creating them and create some alarms if the service exceeds your budget.

There are more than 100 AWS services, and it is important to choose the right service to implement your application based on your requirements. In this book, you will learn to create an AWS account and the required AWS services that allow you to run Python applications. To run and deploy the Python application in AWS, you will learn how to configure the AWS services and deploy them afterward.

Python is also one of the most widely used programming languages. It is easy to learn and has broader usage. Within AWS, most application-related services support Python because of its broad usage, and these services are stable when it comes to the use of Python. AWS always adapts Python use cases with their services, which is a big advantage.

This book is meant for cloud developers, software developers, and IT specialists who want to develop Python applications on AWS as well as learn the concepts of appropriate AWS services for implementing Python applications. You should have Python knowledge, and this book will focus on creating Python applications in AWS. The focus will be on creating and giving details for AWS services instead of digging into Python syntax details. Hence, you will add more expertise to your skillset.

While reading this book, it is important to follow the exercises. This is not just a book of theory and definitions. You will see code examples to illustrate what you have learned. I would recommend implementing the same examples by yourself to help you learn better and apply the same methodologies to your cloud projects. This idea slows down your progress, but you will learn better and easily remember the concepts while using AWS in your professional work life.

At the end of this book, you will implement a graduation project with Python on AWS to connect different AWS services in one application. This project helps you to use different services in the same application and understand the connection between them; you will consolidate your learning with another hands-on exercise.

Once you have created an AWS account, you will be charged according to what usage you have in a month. You always have to be careful what you use and create in AWS. Another point to note is that some AWS services are free for limited usage. Please check the costs before deciding to use any AWS service. Please be aware that you need to pay for AWS costs while doing the exercises. You can check the pricing at this link: `https://aws.amazon.com/pricing/`.

Let's dig into Python programming on AWS.

Installing Python

To install Python, carry out the following steps:

1. Visit the Python download page, `https://www.python.org/downloads/`, and select the right operating system.

2. Download the installation package and run it afterward:

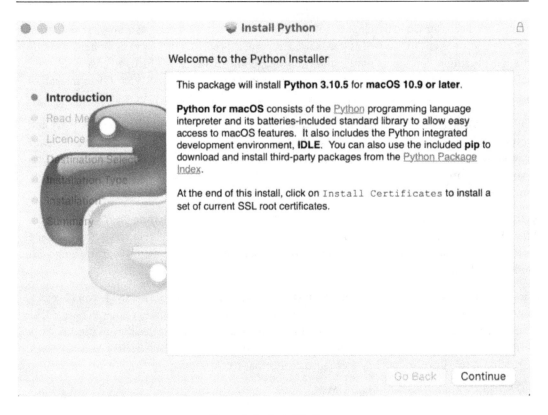

Figure 1.1 – Install Python

After the installation, you will have a Python 3.X folder. The Python folder has the following contents:

Figure 1.2 – Installation folder content

We will follow the steps for macOS; it is very similar to the other operating systems. Let's implement the 'Hello World' application:

1. Double-click on the IDLE application and run the sample 'Hello World' application:

```
                                    IDLE Shell 3.9.11
Python 3.9.11 (v3.9.11:2de452f8bf, Mar 16 2022, 10:34:36)
[Clang 6.0 (clang-600.0.57)] on darwin
Type "help", "copyright", "credits" or "license()" for more information.
>>> print('Hello world')
Hello world
>>> |
```

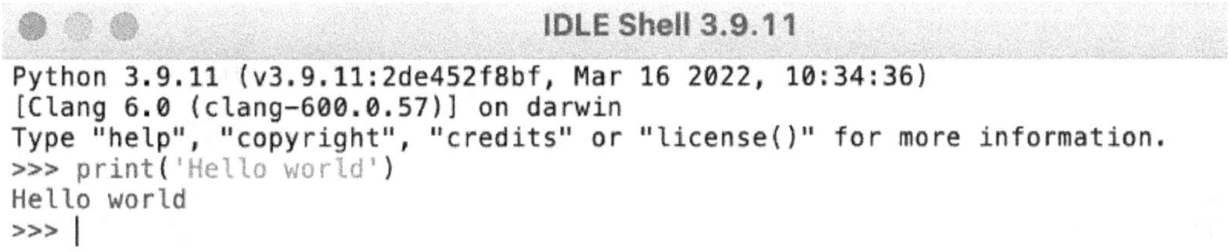

Figure 1.3 – Python command line

If you see this output, congrats! You successfully installed the Python compiler. As a next step, we will install the IDE to simplify the application development.

Installing PyCharm

PyCharm is one of the most powerful IDEs used to develop Python applications. For the examples, we will use PyCharm; you can also use another IDE if you prefer. You have to carry out the following steps:

1. Visit the download page, https://www.jetbrains.com/pycharm/download, and select the right operating system:

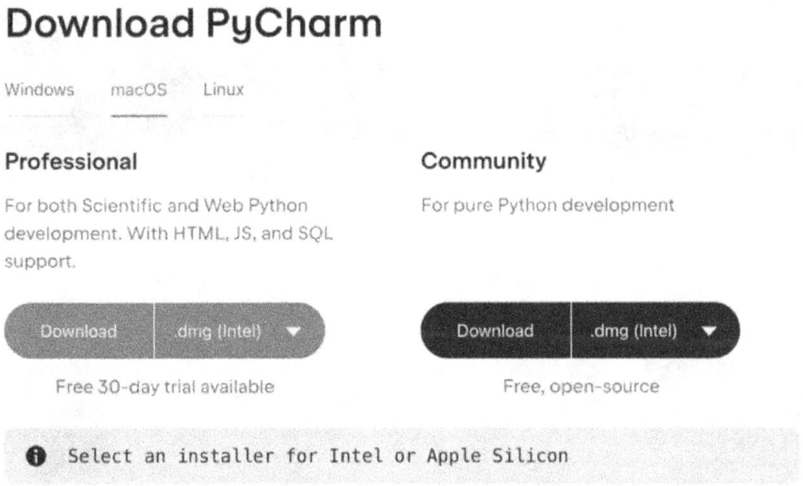

Download PyCharm

Windows   macOS   Linux

Professional

For both Scientific and Web Python development. With HTML, JS, and SQL support.

| Download | .dmg (Intel) ▼ |

Free 30-day trial available

Community

For pure Python development

| Download | .dmg (Intel) ▼ |

Free, open-source

ℹ Select an installer for Intel or Apple Silicon

Figure 1.4 – PyCharm download page

I recommend downloading the **Community** Edition. Otherwise, it will be a trial version for 30 days.

2. Download the installation package and run it afterward. Once you click **Download**, it directly downloads the installation package to the computer:

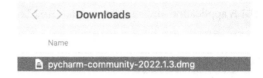

Figure 1.5 – Downloaded folder

When you check the installation folder, you will be able to see the installation program. Install PyCharm onto your machine.

Creating a new project

After the installation of PyCharm, we will create a new project in order to implement our first Python code snippet:

1. Open PyCharm and you will see the **Projects** section:

Figure 1.6 – PyCharm IDE

2. Add a project name:

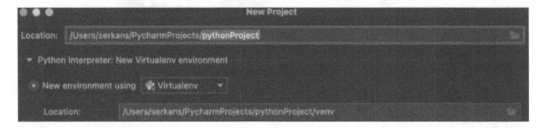

Figure 1.7 – Creating a new project

3. The project is ready to be implemented. Right-click and then click **Run 'main'**:

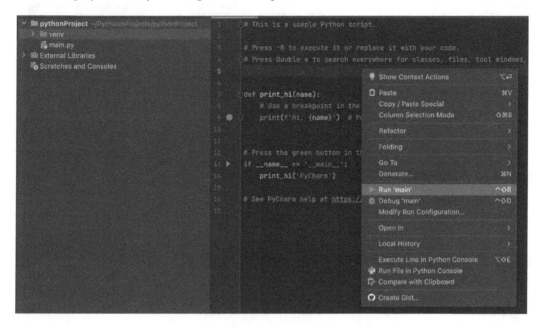

Figure 1.8 – Sample project

4. The command runs the application:

Figure 1.9 – Running the application

Congrats! You have created your first project within PyCharm.

Summary

In this chapter, we explored the cloud basics and advantages. After that, we installed Python and one of the most popular and useful IDEs, PyCharm. PyCharm will be our main tool in order to implement the applications for AWS.

In the next chapter, we will sign up for AWS to have an account on the cloud.

2
Creating an AWS Account

In this chapter, we are going to create an AWS account. This book consists of examples and multiple use cases, so it would be useful to create an account in order to follow along with the exercises in the rest of the chapters on AWS. Let's learn how to create an AWS account.

The chapter covers the following topic:

- Creating an AWS account

Creating an AWS account

To create an AWS account, carry out the following steps:

1. Open the AWS website at https://aws.amazon.com/ in order to create an account.

2. Click the **Create an AWS Account** button on the right side at the top of the page.

Figure 2.1 – The AWS signup page

The **Sign up for AWS** screen will open.

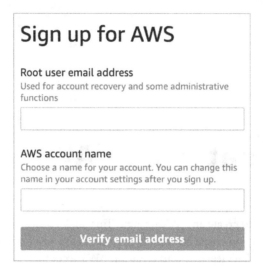

Figure 2.2 – The signup form

3. As can be seen in *Figure 2.2*, there are two fields that need to be completed:

    I.   **Root user email address**: The root user is the owner of all sub-accounts and is able to access all resources and manage them. You can use a single email for the root user. In addition to that, the root user has full access to all services. This is something you need to consider in terms of protecting your account.

    II.   **AWS account name**: The **AWS account name** is an informal name that appears next to the account ID. You can name it while creating an AWS account. You can have multiple accounts under the root account to implement different projects. In some cases, you need to separate the services and costs. In this case, creating multiple accounts could be a good solution.

    Click the **Verify email address** button.

4. Once you fill out the **Root user email address** and **AWS account name** fields, you will receive a verification code via email. This code should be filled out in the **Verification code** input field. Click **Verify**.

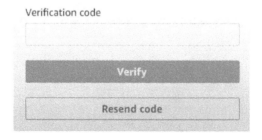

Figure 2.3 – Add the verification code

5. The next step is to define a password for access. Fill out the **Root user password** and **Confirm root user password** fields and click **Continue (step 1 of 5)**.

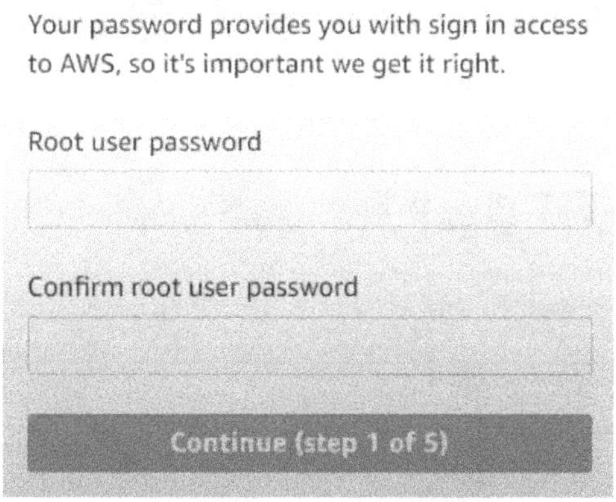

Your password provides you with sign in access
to AWS, so it's important we get it right.

Root user password

Confirm root user password

Continue (step 1 of 5)

Figure 2.4 – Password definition

6. Fill out the personal information required.

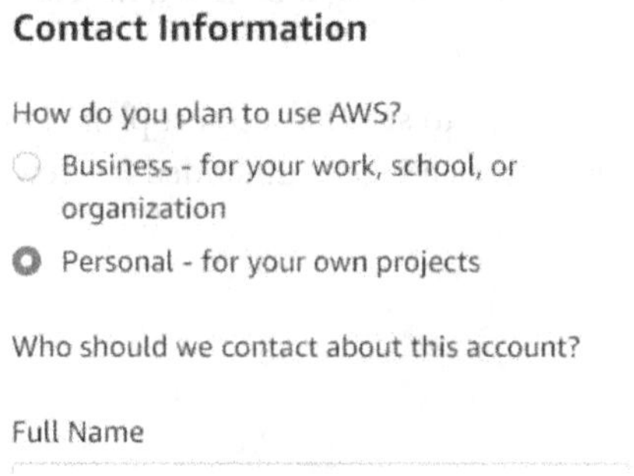

Contact Information

How do you plan to use AWS?

○ Business - for your work, school, or
organization

◉ Personal - for your own projects

Who should we contact about this account?

Full Name

Figure 2.5 – The Contact Information screen

7. After filling out the personal information, fill out the credit card info.

Important note

I would recommend having a budget-limited card, because if you mistakenly open an AWS service that has a big cost or is constantly running, this limited card could prevent you from overspending.

Sign up for AWS

Billing Information

Credit or Debit card number

AWS accepts all major credit and debit cards. To learn more about payment options, review our FAQ

Expiration date

| Month ▼ | Year ▼ |

Cardholder's name

Figure 2.6 – Credit card info

Once you enter the credit card info, you might be asked for confirmation depending on your banking account.

8. After confirming, you will be asked to select a support plan. For learning purposes, you can use the **Basic support - Free** plan, as it is recommended for new users.

Sign up for AWS

Select a support plan

Choose a support plan for your business or personal account. Compare plans and pricing examples. You can change your plan anytime in the AWS Management Console.

- ● Basic support - Free
  - Recommended for new users just getting started with AWS
  - 24x7 self-service access to AWS resources

- ○ Developer support - From $29/month
  - Recommended for developers experimenting with AWS
  - Email access to AWS

- ○ Business support - From $100/month
  - Recommended for running production workloads on AWS
  - 24x7 tech support via email, phone, and

Figure 2.7 – Support plans

Congratulations! After selecting the support plan, you will have an AWS account to get started with the cloud.

Summary

In this chapter, we looked into AWS account creation. The AWS account will help you to carry out Python exercises in the cloud environment. The point to note is that AWS is a paid service and you have to consider the cost of what you are going to use. In the next chapter, we will take a look at popular services such as Lambda.

Part 2:
A Deep Dive into
AWS with Python

In this part, you will deep-dive into the most used AWS services for Python programming, such as Lambda, EC2, and Elastic Beanstalk. However, some other AWS services will be mentioned, such as S3, to gain broader knowledge.

This part has the following chapters:

- *Chapter 3, Cloud Computing with Lambda*
- *Chapter 4, Running Python Applications on EC2*
- *Chapter 5, Running Python Applications with PyCharm*
- *Chapter 6, Deploying Python Applications on Elastic Beanstalk*

3
Cloud Computing with Lambda

In this chapter, we are going to learn the basics of Lambda and implement a Python application to be run in AWS Lambda. For this purpose, we will use our AWS account.

The chapter covers the following topics:

- Cloud computing
- What is Lambda?
- A sample application with Lambda
- Important configurations in Lambda
- A Lambda skeleton
- A Lambda returning value
- Logging in Lambda
- Filing a metadata parser application with Lambda and S3

Cloud computing

Cloud computing allows you to use computer resources such as disk and memory without managing an infrastructure. The concept of the cloud is important in order to free you up to focus on your application. When you use your infrastructure, you need to buy or hire a computer, install all the necessary software, wire the cables, and keep the computer safe from physical as well as soft attacks. It is clear that it takes a significant amount of time; hence, your focus will be on reducing configuration time for your application. With cloud computing, you don't have this kind of headache. The cloud provider takes most of the responsibility and sets up and maintains the data center for you. What you need to do is carry out some configuration and deploy your application to the data center. It makes your life easier; the cloud provider focuses on the infrastructure and you focus on the application. This is the biggest advantage of cloud computing.

What is Lambda?

Lambda is a computing service that allows you to run Python, Java, Node.js, Ruby, .NET, and Go code without provisioning and managing any server. In AWS, it is one of the most used services in the AWS stack. The only thing you need to do is develop and run your code. Lambda also has some advantages in terms of cost.

Lambda is a container that is created by AWS in order to execute your application. When you create a Lambda function, AWS creates this container for you. Hence, you don't need to provision an instance and install the compiler in the container. The only responsibility is to run your code when selecting Lambda.

The advantages of Lambda

The advantages of Lambda are as follows:

- There's no need to provision a server
- It is a pay-as-you-go model
- It supports different runtimes such as Python, Java, and C#
- There's no need to install a software development kit, since it is ready to develop
- It has scalability features – if your process needs more resources, Lambda automatically scales it
- It saves a lot of time for your operational management
- It is able to constantly monitor your Lambda functions

The limitations of Lambda

The limitations of Lambda are as follows:

- **Timeout limit**: If you have long-running functions, Lambda is not the best option. For now, Lambda has a 15-minute timeout limit. If the duration exceeds 15 minutes, you will receive a timeout error.
- **Memory limit**: When you run the function, the process needs memory allocation based on the process flow. If your process needs a massive amount of memory, you will receive an error. In addition to that, Lambda's cost is tied to the execution time and memory used.

You can check the up-to-date limits on the AWS Lambda quotas page: `https://docs.aws.amazon.com/lambda/latest/dg/gettingstarted-limits.html`.

In this section, we looked at some advantages and limitations of Lambda. It is very useful when you need to run any type of application quickly, with no need for a server or detailed installation. Now, we will implement a simple application to learn Lambda and use these advantages to our benefit.

A sample application with Lambda

We are going to execute a sample application within Lambda step by step. To run a Python application on Lambda, take the following steps:

1. Go to the AWS Management Console.

2. Type `lambda` in the search box and click on the Lambda service:

Figure 3.1 – AWS Management Console

3. Click **Create function**.

4. On the **Create function** page, select **Use a blueprint**, and within the blueprint, select the **hello-world-python** application:

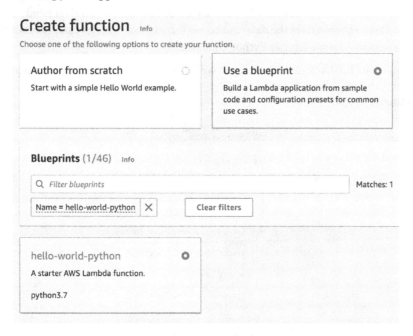

Figure 3.2 – Create function

5.   On the next screen, enter the name of the Lambda function and select the security settings:

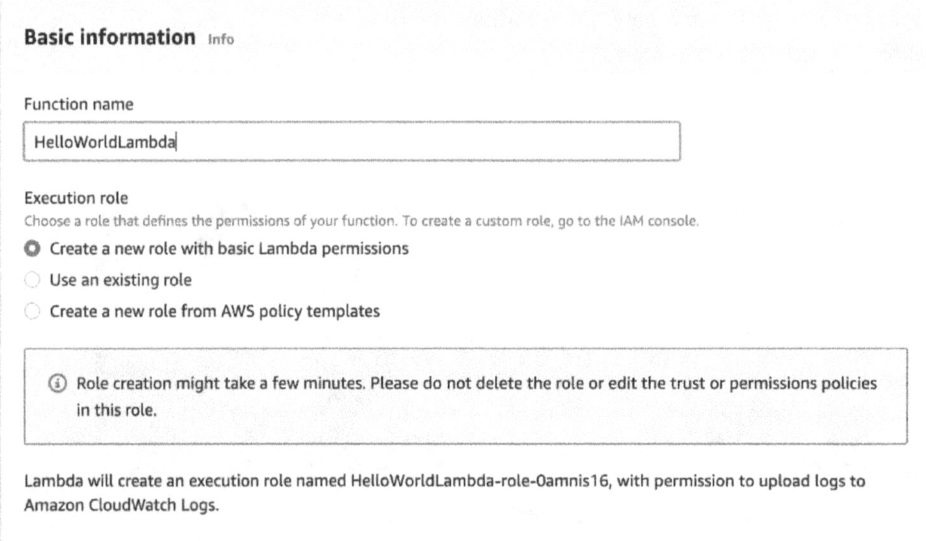

Basic information Info

Function name

HelloWorldLambda

Execution role
Choose a role that defines the permissions of your function. To create a custom role, go to the IAM console.

⦿ Create a new role with basic Lambda permissions

○ Use an existing role

○ Create a new role from AWS policy templates

> ⓘ Role creation might take a few minutes. Please do not delete the role or edit the trust or permissions policies
> in this role.

Lambda will create an execution role named HelloWorldLambda-role-0amnis16, with permission to upload logs to
Amazon CloudWatch Logs.

Figure 3.3 – Naming the function

When you run a Lambda function, you need to define the role that Lambda can use to be able to do some actions, which is done under **Execution role**. The role defines your permissions in AWS and how to access other AWS services. For example, if Lambda needs to access a database, then it should have the database access security role. In this case, Lambda will have basic permission to run a sample Python function.

Once you create the Lambda function, you will have basic Python code to be tested:

Figure 3.4 – A sample Lambda function

6.   Click the **Test** button. When you click it, you can also set the parameters:

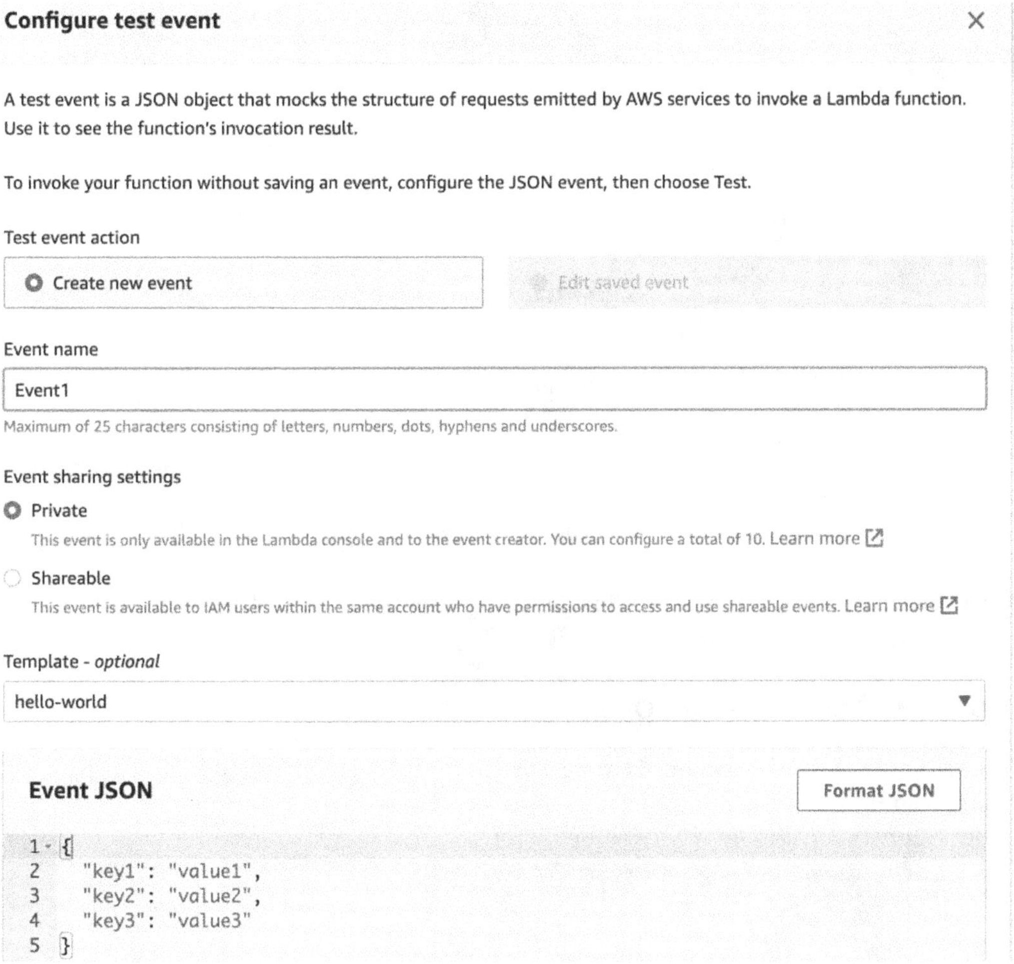

Configure test event   ✕

A test event is a JSON object that mocks the structure of requests emitted by AWS services to invoke a Lambda function. Use it to see the function's invocation result.

To invoke your function without saving an event, configure the JSON event, then choose Test.

Test event action

⦿ Create new event   Edit saved event

Event name

Event1

Maximum of 25 characters consisting of letters, numbers, dots, hyphens and underscores.

Event sharing settings

⦿ Private

This event is only available in the Lambda console and to the event creator. You can configure a total of 10. Learn more ☑

◯ Shareable

This event is available to IAM users within the same account who have permissions to access and use shareable events. Learn more ☑

Template - *optional*

hello-world   ▼

Event JSON   Format JSON

```
1 ▾ {
2       "key1": "value1",
3       "key2": "value2",
4       "key3": "value3"
5   }
```

Figure 3.5 – Running the Lambda function

After running the test, Lambda will run, and you will be able to see the results:

Figure 3.6 – The output of the Lambda function

We have created a sample Lambda function. Once you implement the application, as you can see, running the application is very easy.

Important configurations in Lambda

When you create a Lambda function, there are different configurations that need to be done in order to run it in an efficient way:

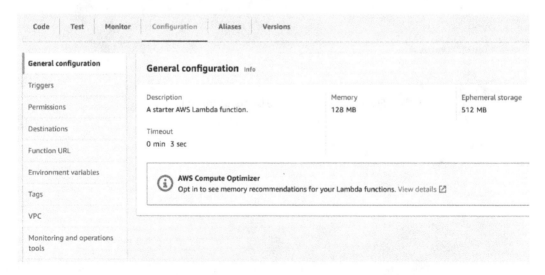

Figure 3.7 – The Lambda configuration

We will use these configurations for the next example. Before starting with the example, let's take a look at the definitions of the configurations:

- **Memory**: This configuration is used to define the memory limit of the application. You need to find the feasible amount of this value. If you define a large amount that is not used, it affects the cost. On the other hand, if you define a smaller amount of memory than is used, your application gives an out-of-memory exception.

- **Timeout**: We mentioned that the Lambda function has a limitation in terms of timeout. You can provide a duration limit under which the Lambda function is supposed to work.

- **Ephemeral storage:** This configuration allows setting a limit for a temporary filesystem. When you run the Lambda application, the /tmp folder is used for temporary storage and needs to be deleted after Lambda finishes the process.

- **Triggers**: Triggers allow you to select an AWS source that runs a Lambda function. For example, S3, an object storage mechanism in AWS, could be a trigger for a Lambda function. We can add S3 configuration in Lambda such that when an object/file is uploaded to S3, it triggers Lambda.

- **Permissions**: Permissions define what roles the Lambda function is able to access. For example, if you need to upload a file to S3 using a Lambda function, then the Lambda function should have an S3 object PUT permission in the execution role.

- **Destinations**: When Lambda finishes the process, it can send information to other services, such as a queue.

- **Environment variable**: This allows you to add an environment variable to be used in a Lambda application. For example, you can add a database URL to this configuration. If the database URL is changed, you don't need to change the code.

- **Tags**: Tags allow you to add a label to your AWS services. It is a good practice for when you search for or categorize services. For example, you may have two similar Lambda functions, the first of which is deployed by the **Customer Relationship Management** (**CRM**) team and the second of which is deployed by the order management team. Hence, you can give two tags to the functions, such as CRM and Order Management, allowing you to categorize your functions and facilitate searching as well. This is also used for cost management.

- **Virtual Private Cloud** (**VPC**): A VPC allows you to create AWS services in a virtual network environment that you define. You can separate AWS services into different network settings. As you see in the following diagram, two instances can be created in different environments:

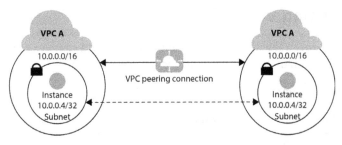

Figure 3.8 – A VPC

- **Monitoring and operations tool**: Lambda collects application logs by default, and they can be monitored via CloudWatch, which helps you to monitor an application. This tool is enabled by default, but you can also disable it.

The configuration of Lambda is important when creating a new function. It is good to know what configuration is used for what reason, hence enabling you to use Lambda in the right way.

A Lambda skeleton

When you implement a Lambda function via Python, you need to follow some rules in order to execute the application. When a Lambda function is run, it calls the `handler` method, which is shown with the following syntax:

```
def lambda_handler(event, context):
    ...
    return some_value
```

As you see, the first parameter is the `event` object. An `event` object consists of JSON in order to process data as a parameter. You can see a sample parameter here:

```
{
   "Temperature": 10,
   "Wind": -5
}
```

The second parameter shows information about the Lambda runtime. You can see some of the runtime fields here:

- `function_name` (the name of the function)
- `function_version` (the version of the function)
- `memory_limit_in_mb` (the Lambda function memory limit)

We've looked at the main skeleton of the Python Lambda function. In the next section, we'll see how to return a value from Lambda.

Lambda returning value

In Lambda, you can return a value that is either a simple message or a complex event with JSON. In the following example, you can see a sample returning message for Lambda:

```
def handler_name(event, context):
    message = 'Weather details. Temperature: {} and Wind: {}!'.
format(event['Temperature'], event['Wind'])
    return message
```

In this example, Lambda takes `Temperature` and `Wind` as input and returns these parameters as a message. In the following example, you can see a more complex return value:

```
def handler_name(event, context):
    return {
      "statusCode": 200,
      "Temperature": 10,
      "Wind": -5
    }
```

As you can see in this example, the return value consists of a simple object to be parsed by the invoker. For example, if Lambda is called by one of the Python applications, this object will be returned once Lambda finishes the process. In general, this parameter allows you to run a Python application with different behavior. In the next section, we'll see how to log information in Lambda.

Logging in Lambda

It is important to use logging functionality in order to trace your application. In some cases, you need to get information about an application; alternatively, you may be processing data via Lambda and you may get an exceptional result. Hence, logging is helpful to check the information to understand the real problem in the application.

There are multiple logging libraries that you can use in Lambda, including this one: `https://docs.python.org/3/library/logging.html`

In the following example, just add a log and return a value:

```
import logging
logger = logging.getLogger()
logger.setLevel(logging.INFO)
def handler_name(event, context):
    logger.info('Process has finished and result will be returned')
    return {
      "statusCode": 200,
      "Temperature": 10,
      "Wind": -5
    }
```

I always recommend adding some logs within an application; it is one of the best practices for being a good developer. In addition to that, we are going to dive deeper into CloudWatch, which is a logging and monitoring service in AWS.

Filing a metadata parser application with Lambda and S3

We are going to execute another application within Lambda. In this case, Lambda will be triggered by S3. S3 is an object storage service to which you can upload different types of files, such as image, CSV, and text files. In this example, when you upload a file to S3, the service will trigger the Lambda function, which in turn will provide information about file metadata. Let's implement the application step by step:

1. Log in to the AWS Management Console.

2. Type lambda in the search box and go to the Lambda service.

3. Click **Create function**.

4. On the **Create function** page, select **Author from scratch** and then in the **Runtime** field, select **Python 3.9**:

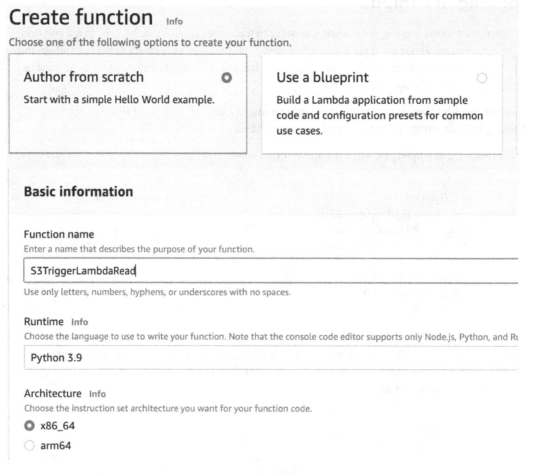

Figure 3.9 – Create function

5. In the **Permissions** section, select **Amazon S3 object read-only permissions** under **Policy templates** and enter a role name. In this case, I entered `S3TriggerLambdaReadRole`. The role is required to read the file from the S3 service:

Permissions   Info

By default, Lambda will create an execution role with permissions to upload logs to Amazon CloudWatch Lo

▼ **Change default execution role**

Execution role
Choose a role that defines the permissions of your function. To create a custom role, go to the IAM console.

○ Create a new role with basic Lambda permissions

○ Use an existing role

● Create a new role from AWS policy templates

>   ⓘ Role creation might take a few minutes. Please do not delete the role or edit the trust

Role name
Enter a name for your new role.

> S3TriggerLambdaReadRole

Use only letters, numbers, hyphens, or underscores with no spaces.

Policy templates - *optional*   Info
Choose one or more policy templates.

> Amazon S3 object read-only permissions ✕
> S3

Figure 3.10 – Permissions

6. Click the **Create function** button at the bottom of the page:

Figure 3.11 – Create function

7. In order to read object metadata, paste the following code snippet into the Lambda function and click the **Deploy** button:

```python
import json
import urllib.parse
import boto3

print('Loading function')

s3 = boto3.client('s3')


def lambda_handler(event, context):
    #print("Received event: " + json.dumps(event, indent=2))

    # Get the object from the event and show its content type
    bucket = event['Records'][0]['s3']['bucket']['name']
    key = urllib.parse.unquote_plus(event['Records'][0]['s3']['object']['key'], encoding='utf-8')
    try:
        response = s3.get_object(Bucket=bucket, Key=key)
        print("CONTENT TYPE: " + response['ContentType'])
        return response['ContentType']
    except Exception as e:
        print(e)
        print('Error getting object {} from bucket {}. Make sure they exist and your zbucket is in the same region as this function.'.format(key, bucket))
        raise e
```

You can also find the original code block from AWS: https://docs.aws.amazon.com/lambda/latest/dg/with-s3-example.html.

Boto3 is used to manage AWS services for Python. We created an S3 client to access and manage the S3 service.

The application is triggered when you put a file into S3. In the code snippet, the code gets the bucket information from the bucket variable. The urllib library allows you to parse an S3 key in order to retrieve an S3 object via the get_object method. Then, we print the content type.

You can also see the latest code snippet within Lambda:

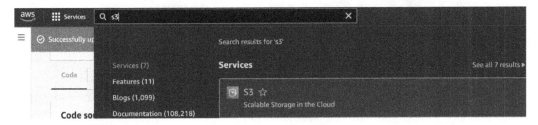

Figure 3.12 – A Lambda function with code

8.   It is time to create an S3 object. Type s3 in the AWS Services search box:

Figure 3.13 – Searching S3

9.   Go to the **S3** service.

10. Within the **S3** service, click the **Create bucket** button:

Buckets (10)   Info
Buckets are containers for data stored in S3. Learn more

Copy ARN   Empty   Delete   **Create bucket**

Figure 3.14 – Creating an S3 bucket

11. Give a unique name to the S3 bucket. The bucket is like a folder, and you can authorize it to upload files such as image and CSV files. Note that the bucket name should be unique:

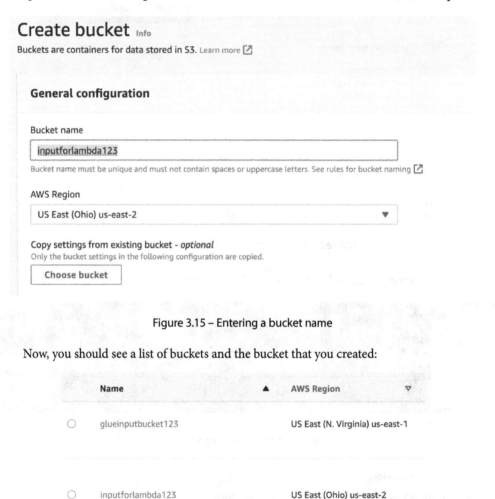

Figure 3.15 – Entering a bucket name

Now, you should see a list of buckets and the bucket that you created:

Name		AWS Region	
○ glueinputbucket123	▲	US East (N. Virginia) us-east-1	▽
○ inputforlambda123		US East (Ohio) us-east-2	

Figure 3.16 – A bucket list

We have created an S3 bucket. Now, we need to make a small configuration that triggers a Lambda function when a file is uploaded to S3:

1. Click the bucket link. For this sample, we need to click **inputforlambda123**. It changes based on the creation name that the user inputted at the beginning:

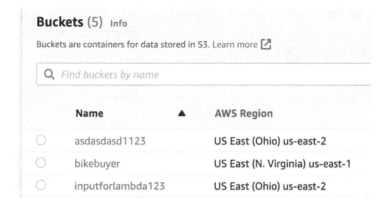

Figure 3.17 – The bucket list

2. Click the **Properties** tab:

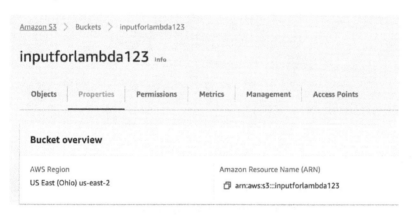

Figure 3.18 – The features of the bucket

3. At the bottom of the **Properties** page, find the **Event notifications** tab.

4. Click the **Create event notification** button:

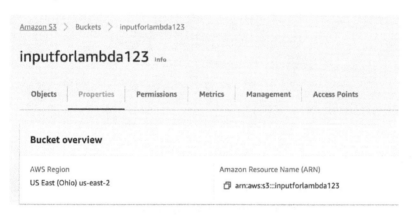

Figure 3.19 – The Event notifications tab

5. In the form, fill out the event name and select the event type in the **Event types** section. For this example, we are going to select the **All object create events** option. Hence, when an object is created, the Lambda function will be triggered:

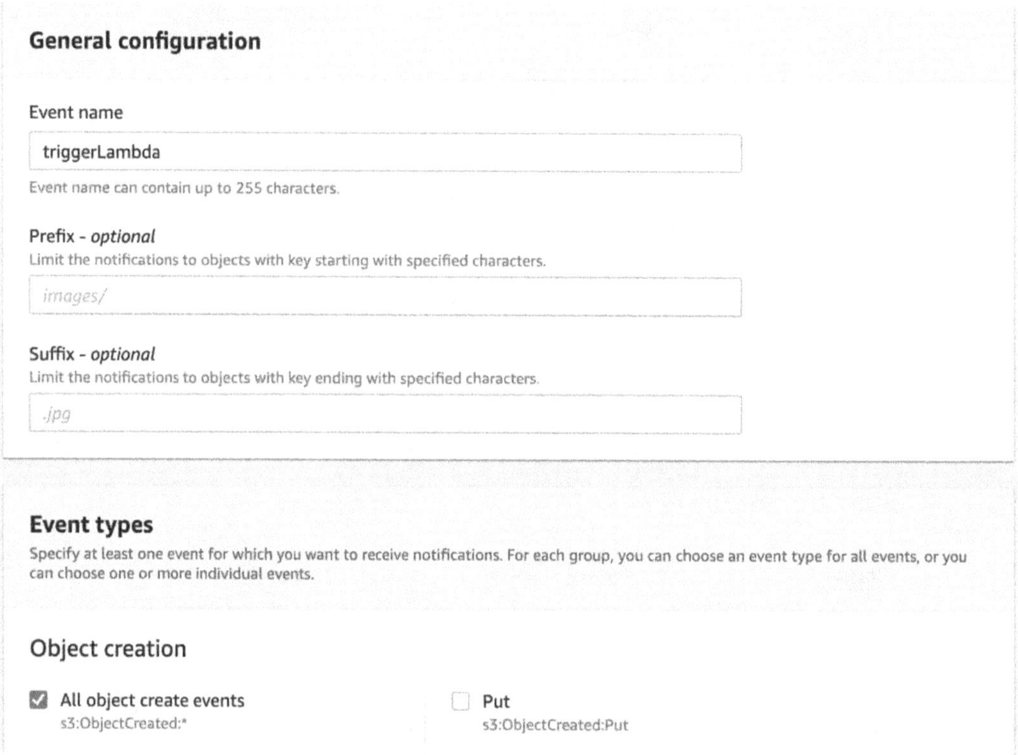

General configuration

Event name

triggerLambda

Event name can contain up to 255 characters.

Prefix - *optional*
Limit the notifications to objects with key starting with specified characters.

images/

Suffix - *optional*
Limit the notifications to objects with key ending with specified characters.

.jpg

Event types

Specify at least one event for which you want to receive notifications. For each group, you can choose an event type for all events, or you can choose one or more individual events.

Object creation

☑ All object create events
    s3:ObjectCreated:*

☐ Put
    s3:ObjectCreated:Put

Figure 3.20 – Event configuration

6. At the bottom of the page, select the Lambda function that will be triggered, under the **Destination** section, and click the **Save changes** button:

Destination

> ⓘ Before Amazon S3 can publish messages to a destination, you must grant the Amazon S3 principal the necessary permissions to call the relevant API to publish messages to an SNS topic, an SQS queue, or a Lambda function. Learn more 🗗

Destination
Choose a destination to publish the event. Learn more 🗗

🔘 **Lambda function**
Run a Lambda function script based on S3 events.

⚪ **SNS topic**
Send notifications to email, SMS, or an HTTP endpoint.

⚪ **SQS queue**
Send notifications to an SQS queue to be read by a server.

Specify Lambda function

🔘 Choose from your Lambda functions

⚪ Enter Lambda function ARN

Lambda function

S3TriggerLambdaRead ▼

Cancel   **Save changes**

Figure 3.21 – The event destination

You should see a success message in the AWS console:

⊘ **Successfully created event notification "triggerLambda".**
Operation successfully completed.

Figure 3.22 – The event destination

You can also double-check with Lambda whether the event has been successfully created. When you click the respective Lambda function, it shows the event source:

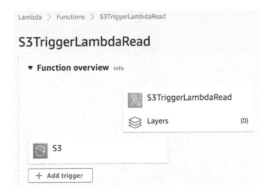

Lambda > Functions > S3TriggerLambdaRead

S3TriggerLambdaRead

▼ **Function overview** Info

🐾 S3TriggerLambdaRead

🥞 Layers   (0)

🗄 S3

+ Add trigger

Figure 3.23 – Lambda with a trigger

At the moment, you are able to see the Lambda function on the left side as a trigger. It is time to test our Lambda trigger:

1. Open the S3 bucket that you created and navigate to it. After that, click the **Upload** button:

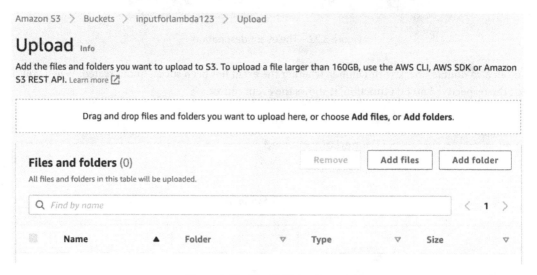

Amazon S3 > Buckets > inputforlambda123

inputforlambda123 Info

Objects	Properties	Permissions	Metrics	Management	Access Points

Objects (0)

Objects are the fundamental entities stored in Amazon S3. You can use Amazon S3 inventory ☑ to get a list of all objec grant them permissions. Learn more ☑

C	🗇 Copy S3 URI	🗇 Copy URL	⬇ Download	Open ☑	Dele

⬆ Upload

Figure 3.24 – An S3 bucket

2. Click the **Add files** button, which allows you to add any kind of file from your computer. For this example, we have uploaded one RTF file. You can also upload an image, PDF, or whatever you want:

Amazon S3 > Buckets > inputforlambda123 > Upload

Upload Info

Add the files and folders you want to upload to S3. To upload a file larger than 160GB, use the AWS CLI, AWS SDK or Amazon S3 REST API. Learn more ☑

Drag and drop files and folders you want to upload here, or choose **Add files**, or **Add folders**.

Files and folders (0)	Remove	Add files	Add folder

All files and folders in this table will be uploaded.

Q Find by name   < 1 >

Name ▲	Folder ▽	Type ▽	Size ▽

Figure 3.25 – The S3 Upload page

The following screenshot shows that you have successfully uploaded the `testTriggr.rtf` file to S3. S3 also gives some details regarding files, such as the type, the latest modification time, as well as the size. If you have more files, you can see a file list under the **Objects** panel:

Figure 3.26 – The S3 file list

As we have uploaded a file to S3, the Lambda function should work. It is time to check whether that is the case. Navigate to the Lambda function:

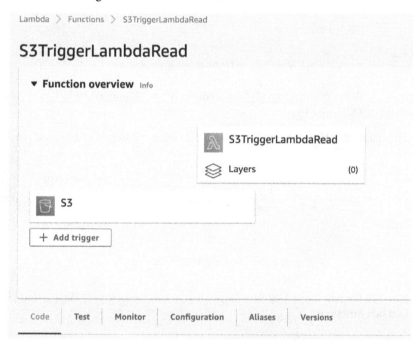

Figure 3.27 – The Lambda function

3. Click the **Monitor** tab, and you should be able to see that the Lambda is called:

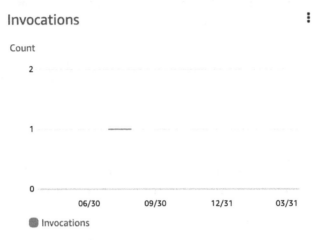

Figure 3.28 — Monitor in Lambda

We can also check the detailed logs via CloudWatch. As we mentioned early, CloudWatch helps you to check AWS service logs.

4. On the same page, click **View logs in CloudWatch**. You will be redirected to the CloudWatch service:

Log stream	Last event time
2022/08/29/[$LATEST]d221e172b11c4e6d989cf3833ab2205d	2022-08-29 16:18:16 (UTC+02:00)
2022/07/28/[$LATEST]859d62e2a7b1441aa818a8da5def217f	2022-07-28 22:42:14 (UTC+02:00)

Figure 3.29 – The CloudWatch service

5. When you click the link under **Log stream**, you will be able to see the logs that you implemented in the Lambda function:

Figure 3.30 – CloudWatch logs

You can also upload different types of files in order to test the Lambda function as well as the CloudWatch logs.

We implemented a simple Python application integrated with S3. When you add a file to a storage mechanism, it triggers the Lambda function in order to process the file. As you saw in this example, you can test your Python code without provisioning a server and installing the Python library. Lambda comes with logging, monitoring, and object storage capabilities.

Summary

In this chapter, we dived into Lambda, which is one of the most important services in AWS. Lambda helps you to deploy and run your application without provisioning a server, which facilitates deployment time. We also touched upon the S3 service, which is used for object storage and has good integration with Lambda. In the following chapter, we will take a look at how to provision a server and run a Python application on an AWS-based server.

4

Running Python Applications on EC2

In this chapter, we are going to learn how to run Python applications within the **Elastic Compute Cloud (EC2)** service. EC2 is an AWS service that allows you to provision a server in the cloud. You can find different types of server options. You need to carry out some configuration and run the server on the cloud. You might wonder why we need EC2 when we have Lambda. Lambda is very effective but has a duration limit. If you run your function for more than 15 minutes, it will give a timeout. What happens if your application needs to be run for a couple of hours because of a huge process? Lambda doesn't work and you need your own server. Another reason to use EC2 would be if you need a very special configuration or installation that needs to be done within a specific server; you would need a server as well. Based on this kind of requirement, you need to have your own server in the cloud. We will provision a server and run a Python application within EC2.

The chapter covers the following topics:

- What is EC2?
- EC2 purchasing options
- EC2 instance types
- Provisioning an EC2 server
- Connecting to an EC2 server
- Running a simple Python application on an EC2 server
- Processing a CSV file with a Python application on an EC2 server
- The AWS CLI

What is EC2?

AWS EC2 is a service that provides a secure and scalable server machine in the cloud. The main advantage of EC2 is that server management is very easy from the AWS Management Console. When you provision an on-premises server, it is not easy to configure security policies, disk management, backup management, and so on. AWS accelerates all this. When you provision EC2, AWS offers different contracts that you need to select and all these types impact the cost.

In order to select the right service, you need to understand what services you are going to use, how many resources you need, and what type of storage you really need. These things are going to help you to reduce the cost and use EC2 efficiently.

EC2 purchasing options

We will now look at the types of EC2 contracts.

On-Demand

In this offer, you don't need to contract for a specific time period. AWS charges according to the time you use the server. You can provision a server, shut it down, and release the server whenever you want. It is a pay-as-you-go model.

Reserved

You need to sign a contract with AWS for 1–3 years. The key thing to note is that AWS offers a discount for a Reserved commitment.

Spot

Let's imagine you have an application that has flexible start and end times. You define a bid price for whatever you are willing to pay for the server. Let's imagine you have a data processing application that runs for five hours and the running time is not important. You are able to run at the beginning or end of the month; it is not a problem. You can provision a Spot instance that significantly reduces your cost.

Dedicated

This is useful when your organization has a software license and is moving to AWS. These servers are only used for your organization. Hence, you can keep the license that is served to your company.

EC2 instance types

AWS offers different types of servers depending on your technical requirement. Server type selection is one of the most important things to manage your budget and use the EC2 server efficiently. If you need to use memory processing applications such as **Spark**, it would be better to provision a memory-optimized server. On the other hand, if you need a server that needs more storage, you can use a storage-optimized server.

The following screenshot shows that you are able to select more than hundreds of types of servers in AWS:

Instance type	vCPUs	Architecture	Memory (GiB)	Storage (GB)	Storage type	Network performance
t2.nano	1	i386, x86_64	0.5	-	-	Low to Moderate
t2.micro	1	i386, x86_64	1	-	-	Low to Moderate
t2.small	1	i386, x86_64	2	-	-	Low to Moderate
t2.medium	2	i386, x86_64	4	-	-	Low to Moderate
t2.large	2	x86_64	8	-	-	Low to Moderate
t2.xlarge	4	x86_64	16	-	-	Moderate
t2.2xlarge	8	x86_64	32	-	-	Moderate
t3.nano	2	x86_64	0.5	-	-	Up to 5 Gigabit

Instance types (498)

Figure 4.1 – EC2 instance types [Source – `https://aws.amazon.com/`]

Auto-scaling

If you need a clustered environment, it would be better to define an auto-scaling policy in order to manage resources efficiently.

Let's think about a batch processing job that runs once a day in order to process massive amounts of data. You provision more than one machine. But when the system is idle, you are going to be charged unnecessarily. However, if you define an auto-scaling policy, the system will close when it is idle. This configuration is going to reduce your costs. The following figure shows the minimum size of the launched instances and the maximum size of the desired capacity:

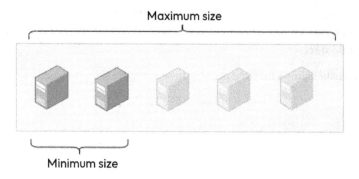

Figure 4.2 – Auto-scaling

Auto-scaling is one of the most important features of EC2. You need to consider the usage of EC2 and configure an auto-scaling feature.

In this section, we took a look at the most important features of EC2. In the next section, we will provision an EC2 server.

Provisioning an EC2 server

We are going to provision an EC2 server step by step. There are different types of EC2 machines; we will provision a free server. I would recommend terminating the server when you finish your work, as we are just using EC2 for learning purposes.

To provision an EC2 server on AWS, carry out the following steps:

1. Go to the AWS Management Console.

2. Search for **EC2** and go to the link titled **EC2**:

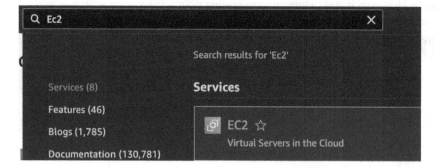

Figure 4.3 – AWS Management Console

3. In order to launch an instance, click **Instances** on the left side, and then click **Launch instances**:

Figure 4.4 – Create an instance

4. In the new panel, you can give a name to the EC2 instance. You can see that we titled ours **Test_Python**. On this launch page, AWS recommends a Linux machine, which is in the free tier. The free tier means that you don't need to pay money to AWS. We will proceed with that option:

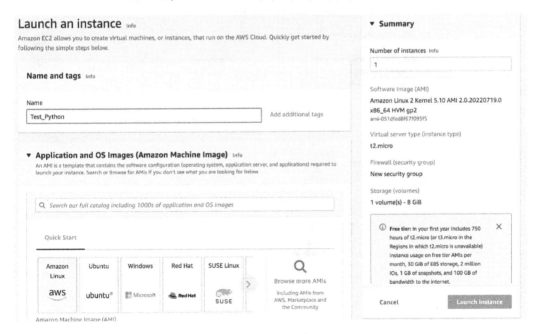

Figure 4.5 – Instance features

5. You can now see the **Key pair (login)** panel. A key pair is used to connect to the server via the SSH key in a secure way. In order to create a new SSH key, click **Create new key pair**:

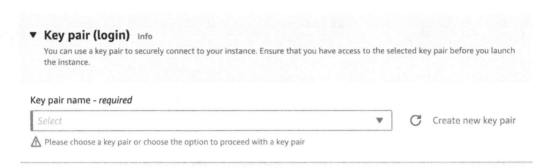

Figure 4.6 – Creating a new key pair

6. We need to give a name to the key pair. Apart from that, you can keep the key pair type and private key file format as the defaults. Click **Create key pair**:

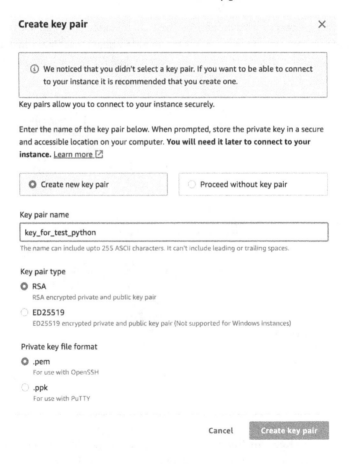

Figure 4.7 – Naming the key pair

Once you click **Create key pair**, it will download the file. Please keep this file; it will be used to connect to the machine. The **Key pair name** dropdown will also be selected with your creation. When you create a new key pair in the upper section, the new key pair name will be visible, which you can see in the following screenshot. For this example, our key pair is **key_for_test_python**:

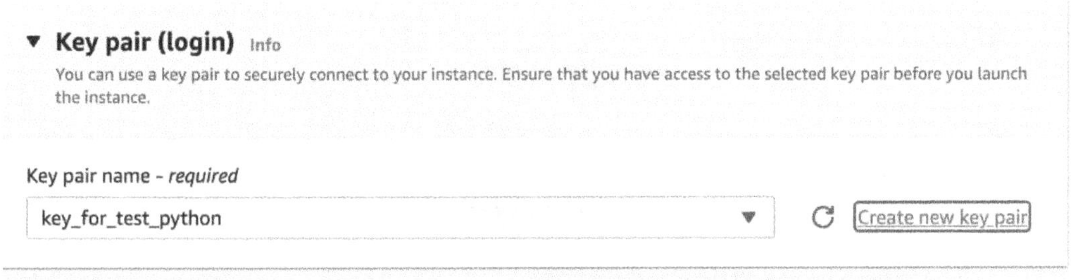

Figure 4.8 – The key pair is ready

In the next step, we are going to create and assign a **virtual private cloud (VPC)** and subnet:

Figure 4.9 – VPC and subnet

A VPC allows AWS services to run in a logically isolated network. It is one of the key services that keep the service secure. You can easily isolate the servers with VPC configuration. The following figure illustrates a VPC and EC2 setup:

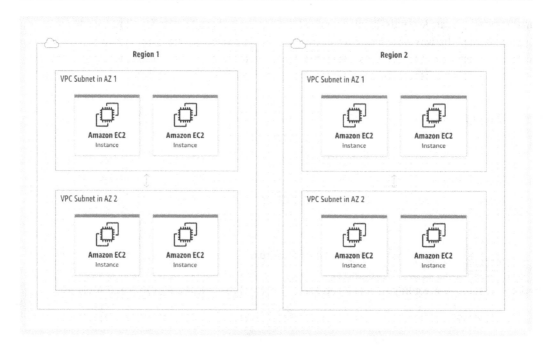

Figure 4.10 – VPC [Source – https://aws.amazon.com/]

As you see, once you add one of the servers to the VPC subnet in AZ 2, it means the EC2 instances are logically isolated from others. Hence, you can add access controls to keep the server secure.

The subnet is also one of the important parts of a VPC. Each VPC consists of a subnet that defines an IP range for the VPC. In the following diagram, you can see the IP range for each subnet:

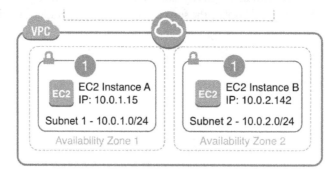

Figure 4.11 – Subnet [Source – https://aws.amazon.com/]

We took a look at VPCs and subnets. Now, we need to define a VPC for the EC2 instance:

1. Type VPC in the search box of the **AWS Management Console**:

Figure 4.12 – VPC on the AWS Management Console

2. Click **Create VPC**:

Figure 4.13 – Create VPC

3. Once you click the button, under the VPC settings, **VPC and more** is selected by default. This option allows you to create a VPC with subnets, which you see on the right side of the following screenshot. With this option, you can create a VPC and subnet together:

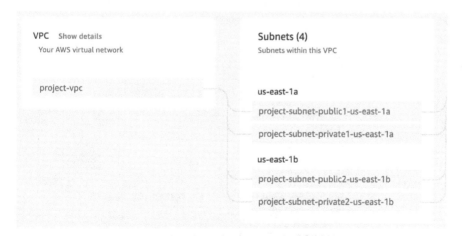

Figure 4.14 – Adding VPC details

4.   At the bottom of this page, click the **Create VPC** button:

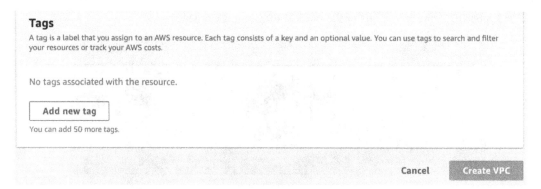

Tags

A tag is a label that you assign to an AWS resource. Each tag consists of a key and an optional value. You can use tags to search and filter your resources or track your AWS costs.

No tags associated with the resource.

Add new tag

You can add 50 more tags.

Cancel   Create VPC

Figure 4.15 – Creating a VPC

When you click **Create VPC**, the VPC begins creation and you can see the status of the progress:

⊘ Success

▼ **Details**

   ⊘ Create VPC: vpc-058e7f6b9a98c8829 ↗

   ⊘ Enable DNS hostnames

   ⊘ Enable DNS resolution

   ⊘ Verifying VPC creation: vpc-058e7f6b9a98c8829 ↗

   ⊘ Create S3 endpoint: vpce-0c6d83ba4bbcce65c ↗

   ⊘ Create subnet: subnet-0e5cd84ea07ea0be3 ↗

   ⊘ Create subnet: subnet-014e1ad14efa7805b ↗

   ⊘ Create subnet: subnet-088b09df85f0155a1 ↗

   ⊘ Create subnet: subnet-00dd3af3cce642264 ↗

   ⊘ Create internet gateway: igw-0687ad5094c223abe ↗

   ⊘ Attach internet gateway to the VPC

   ⊘ Create route table: rtb-0e6e259f011a8f2e4 ↗

   ⊘ Create route

   ⊘ Associate route table

   ⊘ Associate route table

   ⊘ Create route table: rtb-0001f93723753426b ↗

   ⊘ Associate route table

   ⊘ Create route table: rtb-09a68003b281bd08e ↗

   ⊘ Associate route table

   ⊘ Verifying route table creation

   ⊘ Associate S3 endpoint with private subnet route tables: vpce-0c6d83ba4bbcce65c ↗

Figure 4.16 – The VPC creation process

After it has been created, you are able to see the VPC and subnet in the VPC console:

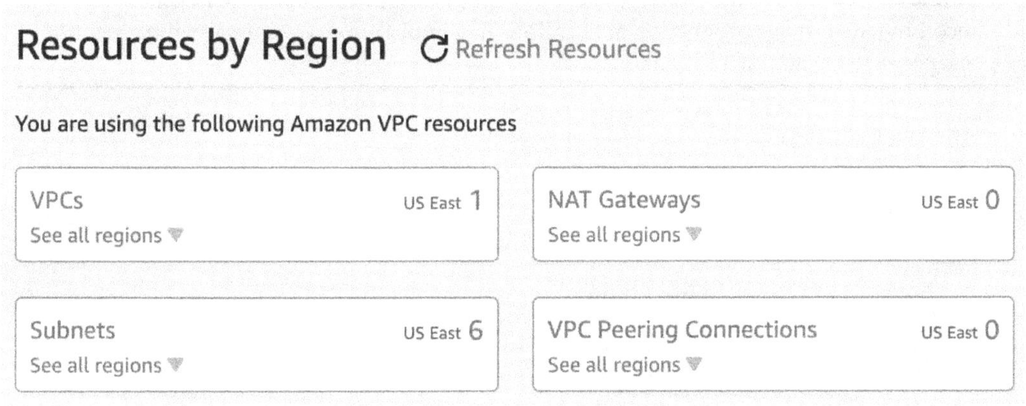

Resources by Region ⟳ Refresh Resources

You are using the following Amazon VPC resources

VPCs	US East 1	NAT Gateways	US East 0
See all regions ▽		See all regions ▽	

Subnets	US East 6	VPC Peering Connections	US East 0
See all regions ▽		See all regions ▽	

Figure 4.17 – The VPC and subnet

So far, we have created a VPC and a subnet. We can proceed with the EC2 creation:

1. Open the **EC2** launch page again. In this case, the VPC and subnet are selected by default. Click **Edit**:

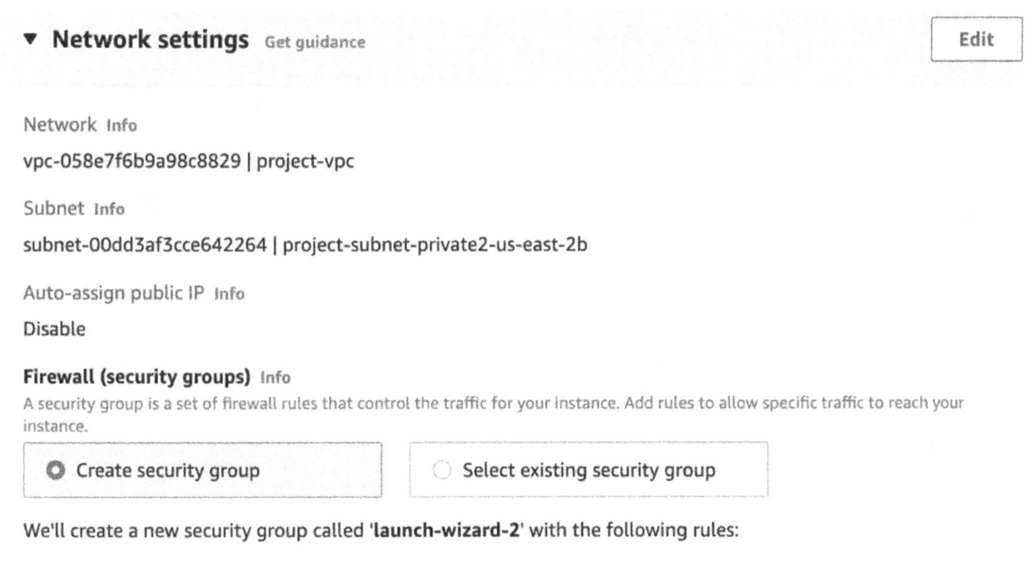

▼ **Network settings** Get guidance   Edit

Network Info

vpc-058e7f6b9a98c8829 | project-vpc

Subnet Info

subnet-00dd3af3cce642264 | project-subnet-private2-us-east-2b

Auto-assign public IP Info

Disable

Firewall (security groups) Info
A security group is a set of firewall rules that control the traffic for your instance. Add rules to allow specific traffic to reach your instance.

● Create security group	○ Select existing security group

We'll create a new security group called '**launch-wizard-2**' with the following rules:

Figure 4.18 – Network settings

2. In order to connect to the machine, we need to select a public subnet and enable **Auto-assign public IP**. You can see the public subnet options in the **Subnet** dropdown. In general, it is not recommended to put production applications in a public subnet. As we are implementing a test project, we can proceed in this manner:

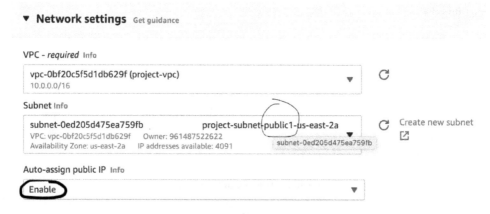

Figure 4.19 – Enabling the public IP

3. At the bottom of the page, click **Launch instance**:

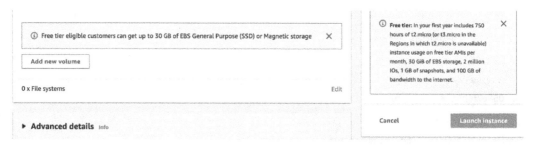

Figure 4.20 – Launching an instance

When we click the **Instances** link on the left side, we are able to see the list of instances that we have created. Congratulations, you have created your first server!

Figure 4.21 – Running instances

You have successfully created a server in an efficient way. We are going to connect to the server in the upcoming section.

Connecting to an EC2 server

In this stage, we are going to connect to the EC2 server via SSH:

1. In the list of instances, there is a **Connect** button. Click it:

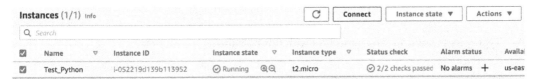

Figure 4.22 – Connecting an instance

2. Under the **SSH client** tab, you can see the steps to connect to the EC2 machine:

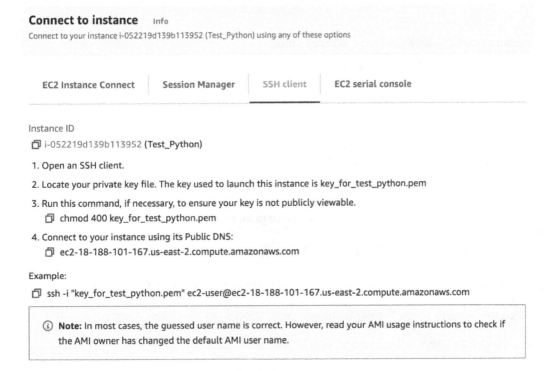

Figure 4.23 – Steps to connect

3. In this example, I will use Mac Terminal in order to connect to the machine via SSH. I am copying the command in the example and pasting it into Terminal. You can also use different SSH applications such as PuTTY and WinSCP. Please make sure the PEM key file is in the same location where you execute the command or that you set the right path for the PEM key file:

```
serkans@NC-serkans-DR49XX4WVR key % ls
key_for_test_python.pem
serkans@NC-serkans-DR49XX4WVR key % ssh -i "key_for_test_python.pem" ec2-user@ec2
-18-188-101-167.us-east-2.compute.amazonaws.com
```

Figure 4.24 – Connecting via Terminal

4. Type `yes` to confirm the connection with this machine:

```
serkans@NC-serkans-DR49XX4WVR key % ssh -i "key_for_test_python.pem" ec2-user@ec2
-18-188-101-167.us-east-2.compute.amazonaws.com
The authenticity of host 'ec2-18-188-101-167.us-east-2.compute.amazonaws.com (18.
188.101.167)' can't be established.
ED25519 key fingerprint is SHA256:BvU4cHBsD4SyC301WIJJXnGo3T+zy6ELIKfqRD8MxRs.
This key is not known by any other names
Are you sure you want to continue connecting (yes/no/[fingerprint])? yes
```

Figure 4.25 – Confirmation for the machine

Congratulations! You have connected to the machine.

```
ED25519) to the list of known hosts.

       __|  __|_  )
       _|  (     /   Amazon Linux 2 AMI
      ___|\___|___|

https://aws.amazon.com/amazon-linux-2/
5 package(s) needed for security, out of 17 available
Run "sudo yum update" to apply all updates.
-bash: warning: setlocale: LC_CTYPE: cannot change locale (UTF-8): No such file o
r directory
[ec2-user@ip-10-0-6-217 ~]$
```

Figure 4.26 – Connected to the machine

You have successfully connected to the server. We are going to install Python in the next section.

Running a simple Python application on an EC2 server

We are going to run a simple Python application on EC2. First of all, check the Python version:

1. Execute `python --version` from the command line:

```
[ec2-user@ip-10-0-6-217 ~]$ python --version
Python 2.7.18
[ec2-user@ip-10-0-6-217 ~]$
```

Figure 4.27 – Checking the Python version

2. Run the `python` command on the command line:

```
[ec2-user@ip-10-0-6-217 ~]$ python
Python 2.7.18 (default, May 25 2022, 14:30:51)
[GCC 7.3.1 20180712 (Red Hat 7.3.1-15)] on linux2
Type "help", "copyright", "credits" or "license" for more information.
>>>
```

Figure 4.28 – Connecting to the Python compiler

3. Run a simple code snippet such as `print 'Hello EC2'` and you will see that the compiler executes the command and prints it:

```
[ec2-user@ip-10-0-6-217 ~]$ python
Python 2.7.18 (default, May 25 2022, 14:30:51)
[GCC 7.3.1 20180712 (Red Hat 7.3.1-15)] on linux2
Type "help", "copyright", "credits" or "license" for more information.
>>> print 'Hello EC2'
Hello EC2
```

Figure 4.29 – Running simple code

We have executed a simple Python application. In the next section, we will run a simple project on EC2.

Processing a CSV file with a Python application on an EC2 server

In the previous chapter, we processed a CSV file within Lambda. In this section, we will run the same application within EC2, but there will be some differences:

1. Log in to the EC2 machine.

2. Create a folder in which to keep the `csv` file that is to be processed.

3. Run the `mkdir csv` command in order to create a `csv` folder on Ubuntu:

```
https://aws.amazon.com/amazon-linux-2/
5 package(s) needed for security, out of 17 available
Run "sudo yum update" to apply all updates.
-bash: warning: setlocale: LC_CTYPE: cannot change locale (UTF-8): No such file
or directory
[ec2-user@ip-10-0-6-217 ~]$
[ec2-user@ip-10-0-6-217 ~]$
[ec2-user@ip-10-0-6-217 ~]$
[ec2-user@ip-10-0-6-217 ~]$ pwd
/home/ec2-user
[ec2-user@ip-10-0-6-217 ~]$ mkdir csv
[ec2-user@ip-10-0-6-217 ~]$ ls
csv
[ec2-user@ip-10-0-6-217 ~]$
```

Figure 4.30 – Creating a folder

After running the `mkdir` command, you can execute with the `ls` command in order to list your directory. As you see, the `csv` folder is created.

4. Locate the `csv` folder by executing `cd csv`:

```
[[ec2-user@ip-10-0-6-217 ~]$ cd csv/
[ec2-user@ip-10-0-6-217 csv]$ 
```

Figure 4.31 – Locating the csv folder

5. Create a sample CSV file in the EC2 machine.

I have uploaded a sample CSV file for you in the following URL. Run the following code to download the sample CSV. The wget command allows you to download the file from the specific link:

```
wget https://raw.githubusercontent.com/PacktPublishing/Python-
Essentials-for-AWS-Cloud-Developers/main/sample.csv
```

```
[ec2-user@ip-10-0-6-217 csv]$ wget https://raw.githubusercontent.com/serkansakinmaz/python-aws-book/main/sample.csv
--2022-08-26 11:30:10--  https://raw.githubusercontent.com/serkansakinmaz/python-aws-book/main/sample.csv
Resolving raw.githubusercontent.com (raw.githubusercontent.com)... 185.199.111.133, 185.199.108.133, 185.199.109.133,
Connecting to raw.githubusercontent.com (raw.githubusercontent.com)|185.199.111.133|:443... connected.
HTTP request sent, awaiting response... 200 OK
Length: 176 [text/plain]
Saving to: 'sample.csv'

100%[===============================================================================>] 176         --.-K/s   in 0s

2022-08-26 11:30:10 (7.77 MB/s) - 'sample.csv' saved [176/176]

[ec2-user@ip-10-0-6-217 csv]$ ls
csvprocess.py  employees.csv  sample.csv
[ec2-user@ip-10-0-6-217 csv]$ cat sample.csv
header 0,header 1,header 2
row 1 col 0,row 1 col 1,row 1 col 2
row 2 col 0,row 2 col 1,row 2 col 2
row 3 col 0,row 3 col 1,row 3 col 2
row 4 col 0,row 4 col 1,row 4 col 2
[ec2-user@ip-10-0-6-217 csv]$ 
```

Figure 4.32 – Downloading the sample CSV file

Now that you have downloaded the file, you are able to create Python code in order to process the CSV file.

6. Run the following code to download the Python code:

```
wget https://raw.githubusercontent.com/PacktPublishing/Python-
Essentials-for-AWS-Cloud-Developers/main/fileprocessor.py
```

```
[ec2-user@ip-10-0-6-217 csv]$ wget https://raw.githubusercontent.com/serkansakinmaz/python-aws-book/main/fileprocessor.py
--2022-08-26 11:36:41--  https://raw.githubusercontent.com/serkansakinmaz/python-aws-book/main/fileprocessor.py
Resolving raw.githubusercontent.com (raw.githubusercontent.com)... 185.199.111.133, 185.199.108.133, 185.199.109.133, ...
Connecting to raw.githubusercontent.com (raw.githubusercontent.com)|185.199.111.133|:443... connected.
HTTP request sent, awaiting response... 200 OK
Length: 272 [text/plain]
Saving to: 'fileprocessor.py'

100%[===============================================================================>] 272         --.-K/s   in 0s

2022-08-26 11:36:42 (16.5 MB/s) - 'fileprocessor.py' saved [272/272]

[[ec2-user@ip-10-0-6-217 csv]$ ls
employees.csv  fileprocessor.py  sample.csv
[ec2-user@ip-10-0-6-217 csv]$ 
```

Figure 4.33 – Downloading the Python code

The following code is very simple; the code imports the csv library and prints the first five lines within the CSV:

```
1   import csv
2
3   with open('sample.csv') as csv_file:
4       csv_reader = csv.reader(csv_file, delimiter=',')
5       line_count = 0
6       for row in csv_reader:
7           print(row)
8           line_count += 1
9           if line_count == 5:
10              break;
11      print('Lines are printed')
```

Figure 4.34 – Python code

7. The next step is to run Python code to see the results. Execute python fileprocessor. py to run the application. After running the application, you will see the results:

```
[ec2-user@ip-10-0-6-217 csv]$ python fileprocessor.py
['header 0', 'header 1', 'header 2 ']
['row 1 col 0', 'row 1 col 1', 'row 1 col 2 ']
['row 2 col 0', 'row 2 col 1', 'row 2 col 2 ']
['row 3 col 0', 'row 3 col 1', 'row 3 col 2 ']
['row 4 col 0', 'row 4 col 1', 'row 4 col 2 ']
Lines are printed
[ec2-user@ip-10-0-6-217 csv]$
```

Figure 4.35 – Running Python code

In this section, we saw how to run a simple Python application within an AWS EC2 server. Now, we will touch upon the AWS SDK for Python.

The AWS CLI

CLI stands for **command-line interface**, which provides some tools and libraries to facilitate accessing AWS services. As such, the AWS CLI has some APIs to use AWS services. The AWS CLI is one of the most common tools used when working with AWS. It has different methods to access AWS services. We are going to install awscli to access AWS services. In this section, we will install awscli and, after that, configure an EC2 machine to upload a file from EC2:

1. In order to access S3 from awscli, we need to create an IAM role to be attached to EC2. Connect to the AWS Management Console, type IAM, and then click **IAM**:

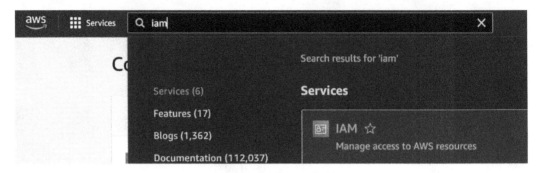

Figure 4.36 – IAM in the console

2.   Click **Roles** on the left panel and then click **Create role**:

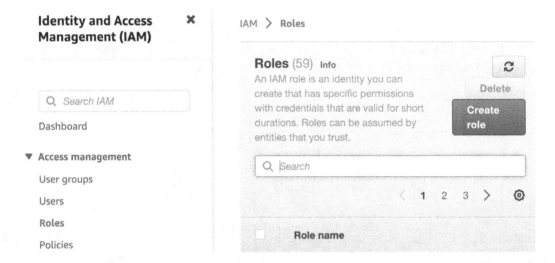

Figure 4.37 – Create role

3.   Select **EC2** as a common use case and click **Next**:

Use case

Allow an AWS service like EC2, Lambda, or others to perform actions in this account.

Common use cases

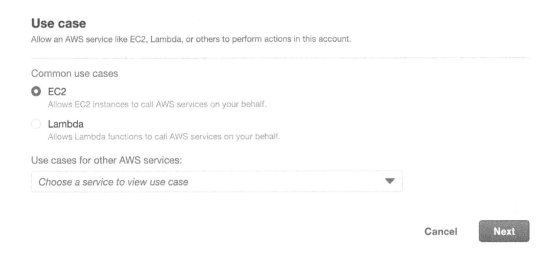

Figure 4.38 – Select a service

4. Now, we need to give the required permission. Since we will access S3, check the **AmazonS3FullAccess** checkbox. This policy will allow users to upload and read the object under S3. After selecting the policy, you can click the **Next** button:

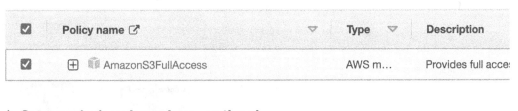

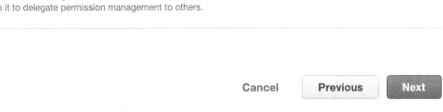

Figure 4.39 – Selecting the policy

5.   Give a name to the role and click the **Create role** button to create a role:

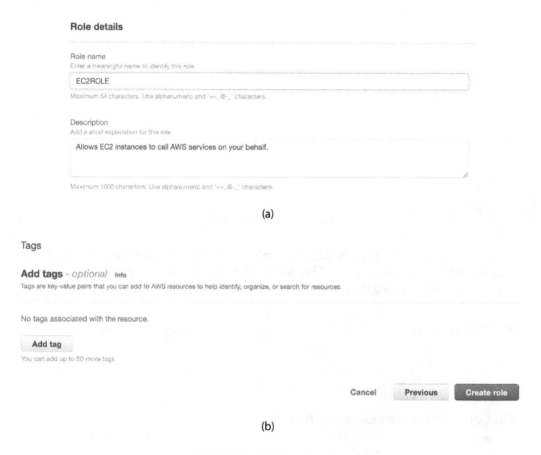

Figure 4.40 – Naming the role

6.   In the final step to attach the role, click the **Actions** drop-down button, go to **Security,** and select **Modify IAM role**:

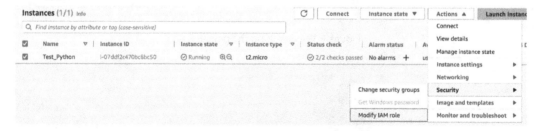

Figure 4.41 – Attach role

7.   On the next screen, select **EC2ROLE**, to be attached to EC2, and click **Update IAM role**:

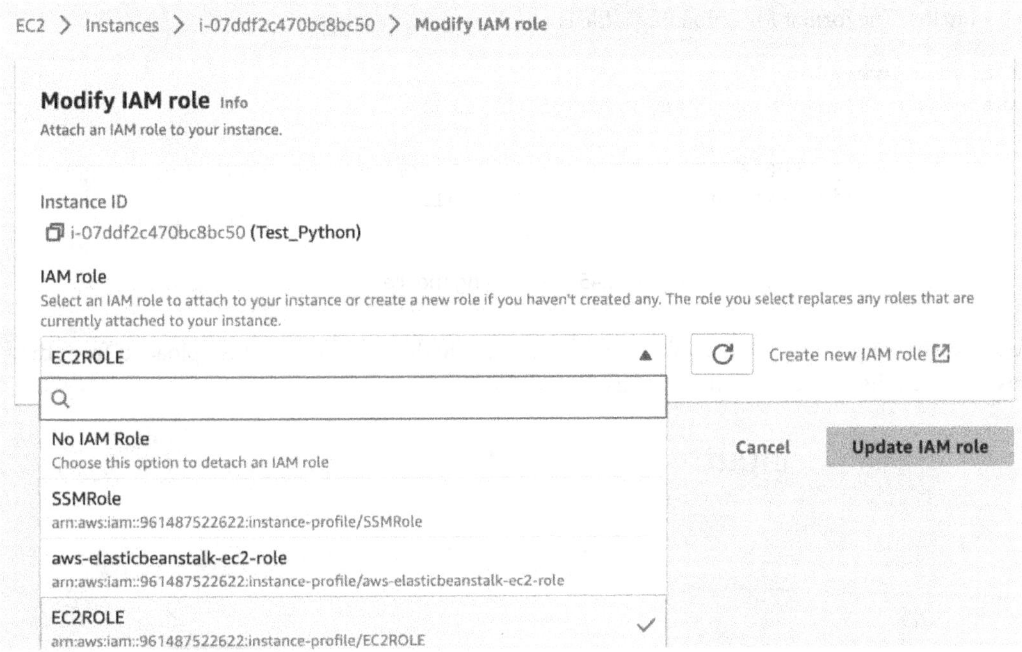

Figure 4.42 – Update IAM role

We have created and attached the required role to log in to an EC2 machine:

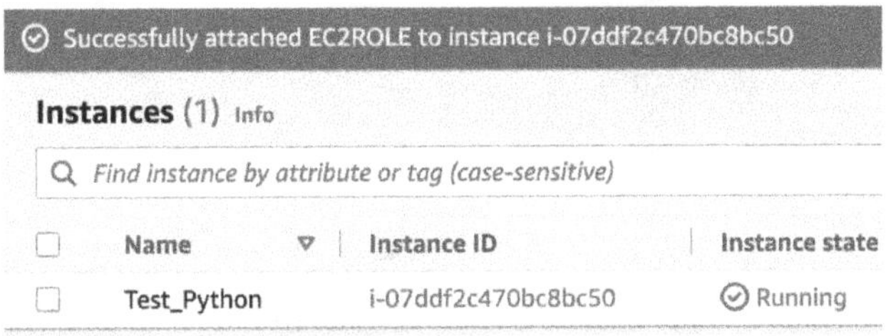

Figure 4.43 – Task complete

Now, we will upload a file to S3.

Create a file under the EC2 machine. The touch command helps you to create an empty file. Optionally, you can also create a file using another application:

```
[ec2-user@ip-10-0-6-217 ~]$ touch file1.txt
[ec2-user@ip-10-0-6-217 ~]$ ls
csv  file.txt  file1.txt
```

Figure 4.44 – Creating a file

We can upload this file to S3 via the AWS CLI. In the previous chapter, we created an S3 bucket. You can use this bucket or create a new bucket to test the AWS CLI S3 command. Let's upload the file to the S3 bucket. The format for uploading a file is as follows:

```
Format : aws s3 cp from to
aws s3 cp file.txt s3://inputforlambda123
```

```
[[ec2-user@ip-10-0-6-217 ~]$ aws s3 cp file.txt s3://inputforlambda123
upload: ./file.txt to s3://inputforlambda123/file.txt
```

Figure 4.45 – Uploading the file

We successfully uploaded the file. We are able to check whether the S3 bucket is uploaded from the console. Open the bucket from the S3 console and check:

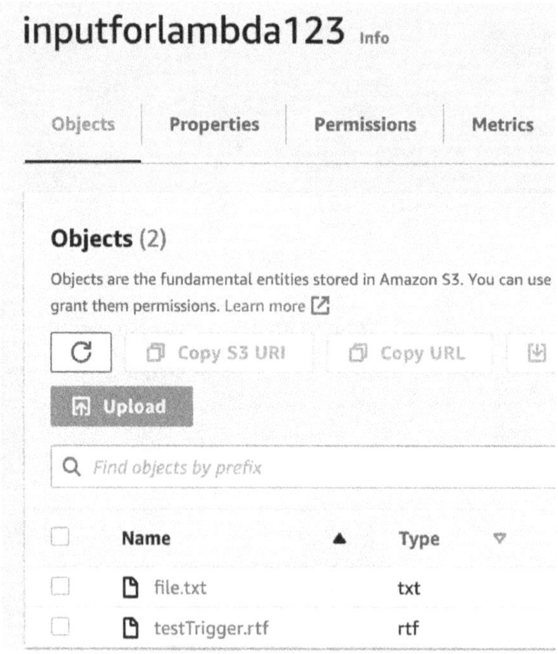

Figure 4.46 – Bucket content

As you can see, the file is uploaded to the S3 bucket.

The AWS client is useful when you want to access AWS services and perform some tasks using commands. In this section, we learned how to copy a file to the S3 bucket via the command line, which saves a lot of time.

Summary

In this chapter, we learned about the AWS EC2 service, which is used to create a server on the cloud. You can create your server in an efficient way and use it for different purposes, such as an application server, web server, or database server. We also created an EC2 server as an example and ran our Python application on EC2. In the following chapter, we will take a look at how to debug our Python application via PyCharm.

5

Running Python Applications with PyCharm

In this chapter, we are going to run a Lambda application with PyCharm. Running Lambda applications via PyCharm is both useful and practical during development as it consists of a code editor, debugger, and common development tools with a developer-friendly graphical user interface. These features of PyCharm help us to easily find bugs in our code.

This chapter covers the following topics:

- Installing the AWS Toolkit
- Configuring the AWS Toolkit
- Creating a sample Lambda function in AWS
- Running an AWS Lambda function using the AWS Toolkit

Installing the AWS Toolkit

In this section, we will install the AWS Toolkit in PyCharm. The AWS Toolkit is an extension for PyCharm to develop, debug, and deploy your applications for AWS. Let's get to it:

1. Open PyCharm on your computer.
2. Open **Preferences** from the **PyCharm** dropdown and select **Plugins**:

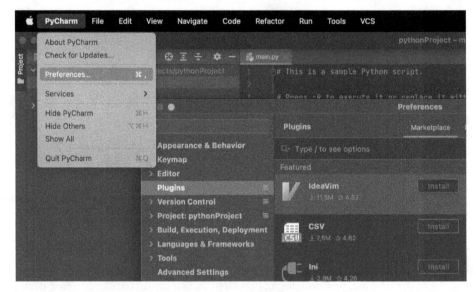

Figure 5.1 – Preferences

3.   Type AWS Toolkit in the search area and click **Install**:

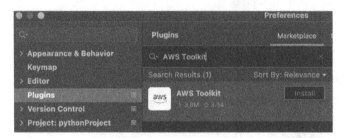

Figure 5.2 — Install the AWS Toolkit

4.   After installation, the IDE will ask you to restart it. Click the **Restart IDE** button:

Figure 5.3 – Restart the IDE

We have installed the AWS Toolkit in PyCharm. As the next step, we are going to configure the credentials for our AWS account.

Configuring the AWS Toolkit

We are going to configure the AWS Toolkit in order to connect it to our AWS account. We will start by setting the credentials for our AWS account:

1. After restarting the IDE, you will see the text **AWS: No credentials selected** at the bottom-right of the page. Click this text:

Figure 5.4 – AWS: No credentials selected

2. After clicking it, you will see the **AWS Connection Settings** menu appear. We are now going to configure the credentials. In order for the IDE to connect to AWS, we need to provide the AWS access key and secret key:

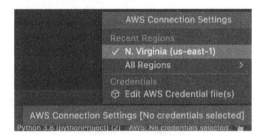

Figure 5.5 – Click Region

In the previous chapter, *Chapter 4, Running Python Applications on EC2*, we created an S3User via the IAM service. For our current use case, we follow the same steps in order to create a user that has Lambda access:

1. In the IAM console, add a user with the name ProgrammaticUser and click **Next: Permissions**:

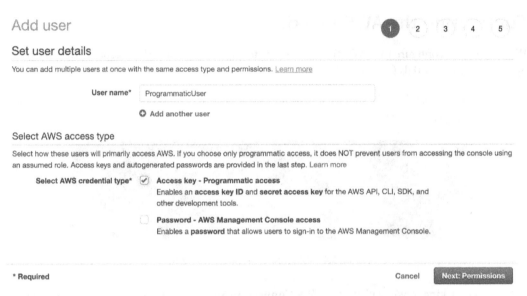

Figure 5.6 – Add user

2.   In the next panel, select **AWSLambda_FullAccess** and proceed to create a new user. The steps are the same as those we used to create the user in the previous chapter. Click **Next: Tags** and proceed:

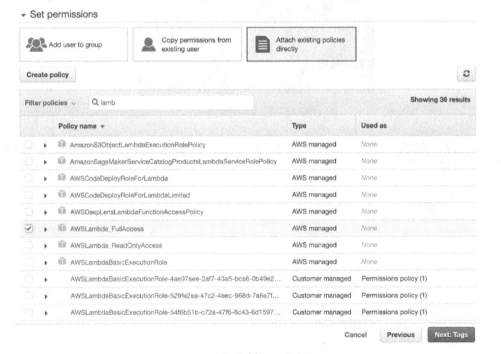

Figure 5.7 – Add permission

3. We will now provide the access key ID and secret access key for the AWS connection setup. Open PyCharm again and click **Edit AWS Credential file(s)**:

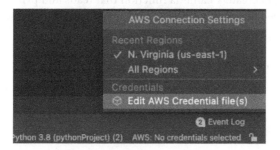

Figure 5.8 – Edit credentials

4. Click the **Create** button on the following dialog window that appears:

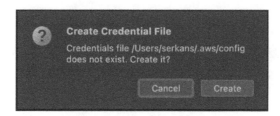

Figure 5.9 – Create the credential file

5. Once you click **Create**, you will be presented with a file in which you can enter the credentials. Place the access key ID and secret access key in the file and save it:

```
[default]
# The access key and secret key pair identify your account and grant access to AWS.
aws_access_key_id = AKIA57XI
# Treat your secret key like a password. Never share your secret key with anyone. Do
# not post it in online forums, or store it in a source control system. If your secret
# key is ever disclosed, immediately use IAM to delete the access key and secret key
# and create a new key pair. Then, update this file with the replacement key details.
aws_secret_access_key = AYJyEeHrJTUB/P9YUB7HQdz
```

Figure 5.10 – Edit the credential file

We have created the AWS credentials and adjusted them in the PyCharm. As a next step, we are ready to create a Lambda function.

Creating a sample Lambda function in AWS

In this step, we are going to create a Lambda function that reads and prints a file from S3. In the previous chapter, we learned how to create the S3 bucket and Lambda function. Hence, we keep the explanation short here:

1. We are going to copy a sample file to the S3 bucket:

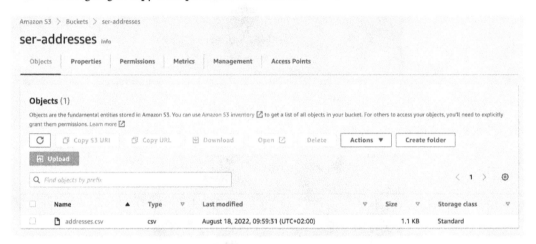

Figure 5.11 – File in S3

2. Create a Lambda function that reads the file from S3. I've called the Lambda function `FileProcessing`; however, you can give it any name that you prefer:

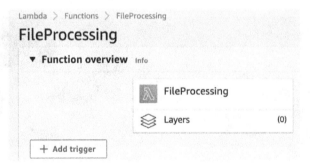

Figure 5.12 – Lambda function

3. Once the Lambda is created, we paste the code to Lambda from the GitHub link under the image. In the code block, we are going to implement a simple function to read the content of the S3 bucket and print it. You can retrieve the code block from the GitHub page that I have shared after *Figure 5.13*. Broadly speaking, the `s3.get_object` method reads the file with

the given parameters of `bucket` and `key`. Once you have a file stored in S3, the content is under the `Body` JSON file and the final step is to print the content:

Figure 5.13 – Code in Lambda

The following GitHub link consists of the code block for the S3 Reader application: `https://github.com/PacktPublishing/Python-Essentials-for-AWS-Cloud-Developers/blob/main/S3Reader.py`.

4.   Click the **Test** button in order to check whether the Lambda function is running. When you click the **Test** button the first time, you need to configure the sample event:

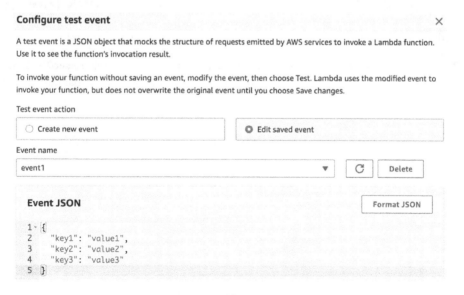

(a)

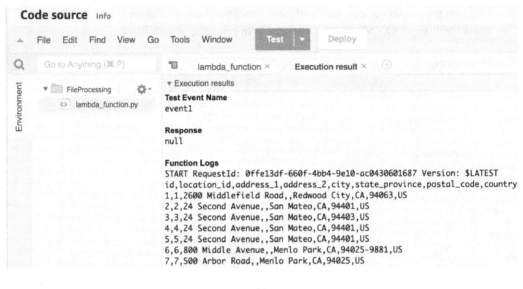

(b)

Figure 5.14 – Configure and test the Lambda function

We have created the Lambda function. In the next section, we are going to run this function within PyCharm via the AWS Toolkit.

Running an AWS Lambda function using the AWS Toolkit

In this section, we are going to run our Lambda function within PyCharm. Let's follow the steps:

1. Open **AWS Toolkit** on the left side of PyCharm and you will be able to see the Lambda functions that are defined in the AWS Lambda service. Seeing this means that the connection we configured works:

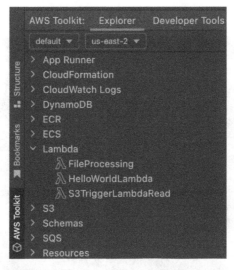

Figure 5.15 – Open the AWS Toolkit menu

In the list, we can see the functions that we created in the **us-east-2** region. We are now ready to run the Lambda function that we created in the previous section.

2. Right-click **FileProcessing** and, on the resulting menu, click the **Run '[Remote] FileProcess...'** button:

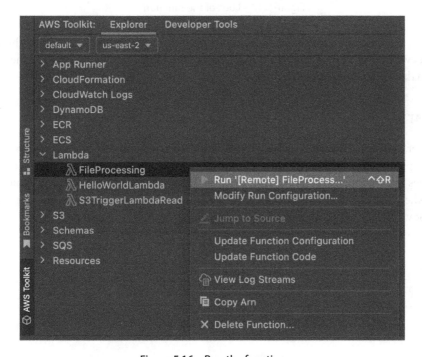

Figure 5.16 – Run the function

When you click the link, the AWS Toolkit will run the Lambda function via PyCharm:

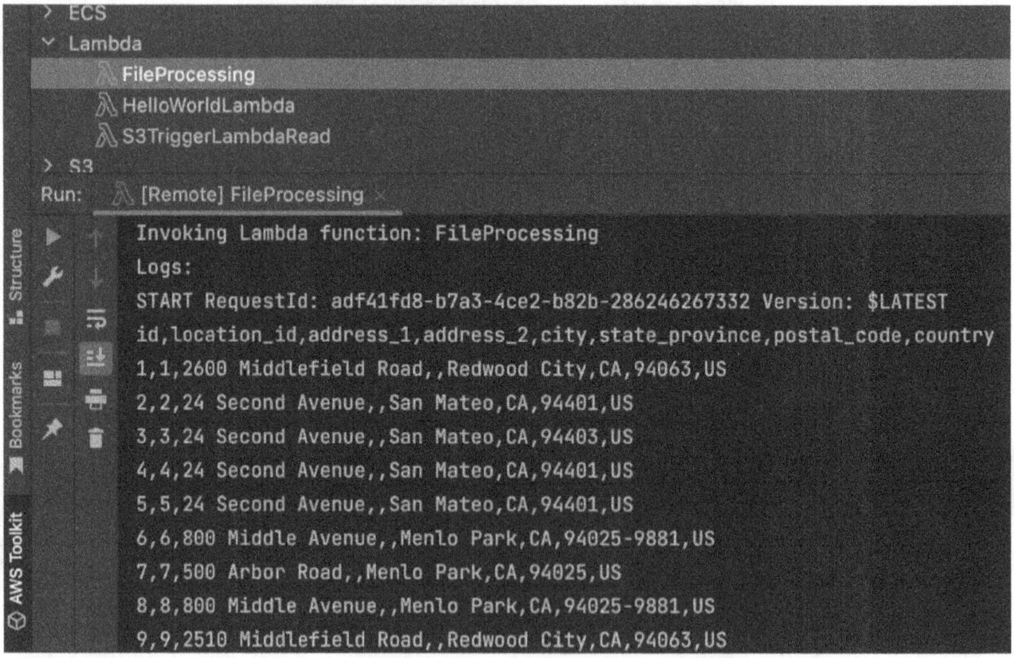

```
>  ECS
∨  Lambda
      λ  FileProcessing
      λ  HelloWorldLambda
      λ  S3TriggerLambdaRead
>  S3
Run:     λ  [Remote] FileProcessing ×

      Invoking Lambda function: FileProcessing
      Logs:
      START RequestId: adf41fd8-b7a3-4ce2-b82b-286246267332 Version: $LATEST
      id,location_id,address_1,address_2,city,state_province,postal_code,country
      1,1,2600 Middlefield Road,,Redwood City,CA,94063,US
      2,2,24 Second Avenue,,San Mateo,CA,94401,US
      3,3,24 Second Avenue,,San Mateo,CA,94403,US
      4,4,24 Second Avenue,,San Mateo,CA,94401,US
      5,5,24 Second Avenue,,San Mateo,CA,94401,US
      6,6,800 Middle Avenue,,Menlo Park,CA,94025-9881,US
      7,7,500 Arbor Road,,Menlo Park,CA,94025,US
      8,8,800 Middle Avenue,,Menlo Park,CA,94025-9881,US
      9,9,2510 Middlefield Road,,Redwood City,CA,94063,US
```

Figure 5.17 – Logs of the function

After running the function, some Lambda logs will appear in PyCharm. As you can see, this makes it easier to develop Python applications for AWS. You can test this from your local machine without logging into the AWS Management Console.

Summary

In this chapter, we learned how to install and use the AWS Toolkit within PyCharm. It is always helpful when you implement and deploy AWS services within PyCharm in a practical way. AWS Toolkit has AWS services integration; therefore, instead of using the AWS Management Console, you can use PyCharm where it is installed on the local machine. In the following chapter, we will take a look at how to deploy a Python application to Elastic Beanstalk.

6

Deploying Python Applications on Elastic Beanstalk

In this chapter, we are going to learn how to deploy Python applications on **Elastic Beanstalk**. Elastic Beanstalk is an AWS service that allows you to deploy web applications in the cloud. Basically, you don't need to provision a server; Elastic Beanstalk provisions an infrastructure in the backend and deploys your web application. Another advantage of Elastic Beanstalk is being able to scale up your web applications when there are a large number of requests from the user.

This chapter covers the following topics:

- What is Elastic Beanstalk?
- Creating a Python web application
- Deploying a simple Python web application on Elastic Beanstalk

What is Elastic Beanstalk?

Elastic Beanstalk is an AWS service that is used to deploy web applications in the cloud. It supports multiple web application frameworks such as Python, Java, .NET, PHP, Node.js, Ruby, and Go. Once you deploy your application, Elastic Beanstalk manages the infrastructure in order to deploy, run, scale, and monitor applications.

Features of Elastic Beanstalk

Let's take a look at the high-level features of Elastic Beanstalk:

- It supports monitoring and logging; hence, you can easily track how the application is behaving. For example, if an application goes down, you can check via Elastic Beanstalk.

- It manages updates for infrastructure. In some cases, your application should be updated with the latest improvements in Python or other libraries and Elastic Beanstalk manages the updates with you in control.

- It manages scaling features up and scaling features down; hence, if your application has too many requests, it adds more resources, and your application can then meet the requests. On the other hand, if there is less demand, it reduces the resources and helps to reduce the cost.

- It supports some financial data or protected health information standards; hence, you can use Elastic Beanstalk for financial applications as well as health information applications.

We have taken a look at the basic features of Elastic Beanstalk, and we will now start to implement a sample web application with Python to deploy via Elastic Beanstalk.

Creating a Python web application

We are going to create a sample web application with Python. For that purpose, **Flask** will be used as a web application framework for Python.

Flask is a web application framework that is written with Python. It has the required libraries to start implementing web applications as a beginner. In the following code block, you can see a sample "**Hello, World!**" web application with Flask:

```
from flask import Flask
app = Flask(__name__)

@app.route('/')
def hello_world():
    return 'Hello, World!'
```

The code imports the Flask library and runs the application on localhost port 5000. When you run it, you will see "**Hello World!**" in the browser.

You can also check the Flask framework at the following website: https://flask.palletsprojects.com/en/2.2.x/.

As the next step, we are going to deploy a Python web application to Elastic Beanstalk.

Deploying a Python web application on Elastic Beanstalk

In this section, we are going to deploy a sample Python web application on Elastic Beanstalk:

1. Type Elastic Beanstalk in the AWS Management Console search box and click **Elastic Beanstalk**:

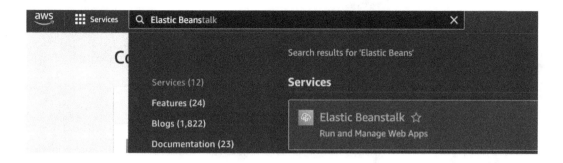

Figure 6.1 – AWS Console

You will see the main page of Elastic Beanstalk:

Figure 6.2 – Elastic Beanstalk

2. Click **Environments** on the left side in order to create a new Python web application, and then click the **Create a new environment** button:

Figure 6.3 – Environment list

3. In the next panel, we are going to select what type of environment we want. Since we would like to deploy a web application, select **Web server environment**:

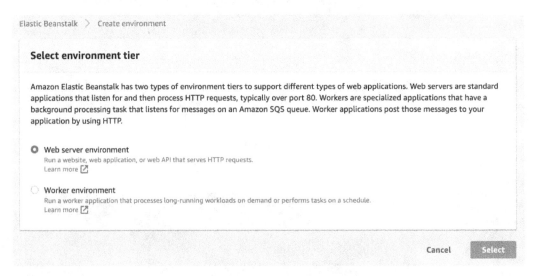

Figure 6.4 – Selecting an environment

4. I have named the file `Python Web` app. You can name it whatever you want:

Create a web server environment

Launch an environment with a sample application or your own code. By creating an environment, you allow Amazon Elastic Beanstalk to manage Amazon Web Services resources and permissions on your behalf. **Learn more** ☑

Application information

Application name

```
Python Web app
```

Up to 100 Unicode characters, not including forward slash (/).

▶ **Application tags (optional)**

Environment information

Choose the name, subdomain, and description for your environment. These cannot be changed later.

Environment name

Figure 6.5 – Naming the application

5. After naming the application, scroll down and fill in the **Environment name** input field. Keep in mind that this can also be named by the AWS Console by default. You have the option to change it.

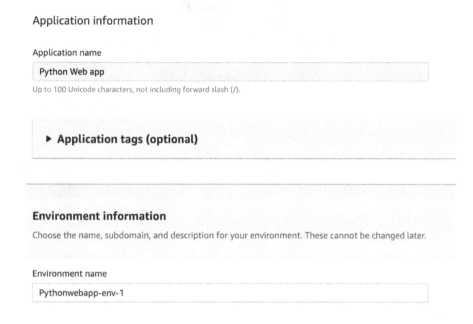

Figure 6.6 – Environment name field

6. When you scroll down further, there is another input field to fill out – **Domain**. The domain will be used to access your web application via the browser. In this example, we will enter test-training and check the availability by clicking the **Check availability** button:

Figure 6.7 – Naming the domain

7. Once you find the available domain name, scroll down, and locate the **Platform** panel. In this panel, we need to select the web application framework. Elastic Beanstalk supports different web environments such as Java, PHP, Node.js, Python, and so on. We will select the Python platform to deploy a Python web application. Depending on which Python platform you are working on, you can select it from the **Platform branch** field. In this example, I am selecting the **Python 3.8 running on 64bit Amazon Linux 2** version. **Platform version** consists of some updates and patches according to the platform. You can proceed with the latest version; for example, if AWS finds a security patch, it creates a new version:

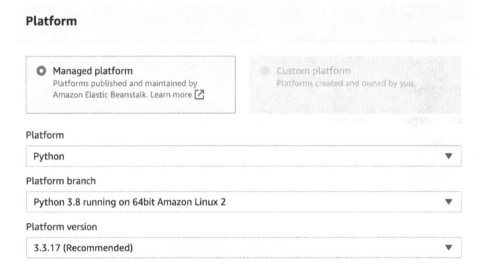

Figure 6.8 – Selecting the platform

8. Scroll down and you will see the latest panel on the page. In this example, we will proceed with **Sample application** and click **Create environment**:

Figure 6.9 – Finalizing the platform

9. Once you click **Create environment**, you will see the logs. Elastic Beanstalk creates the platform and deploys sample applications:

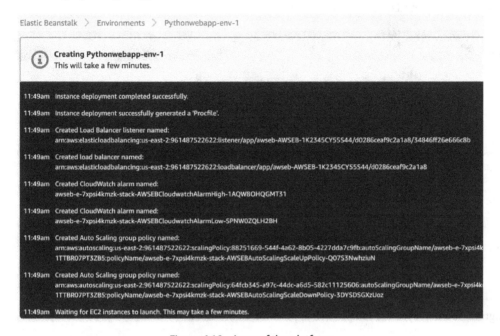

Figure 6.10 – Logs of the platform

Wait a few minutes so that the application is deployed. Once deployed, you will be presented with the following screen:

Figure 6.11 – Application deployment

It seems like the sample application has been deployed and is running properly. Click the domain link to see the running application. In the preceding screenshot, the domain link is `test-training.us-east-2.elasticbeanstalk.com`:

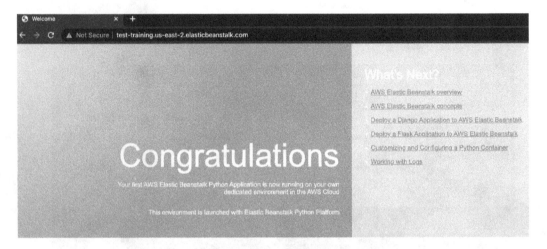

Figure 6.12 – Application

Congrats! You deployed the sample web application to the cloud.

In this example, we deployed the sample application to Elastic Beanstalk. The sample web application is implemented by AWS. As the next step, we are going to implement a simple Python web application to be deployed by Elastic Beanstalk:

1. Open the Elastic Beanstalk service in AWS.

2. Click **Environments** on the left side and see the list of environments. In the previous section, we created an environment and deployed the sample application. In this example, we will use the same Python web environment:

Figure 6.13 – Environments

3. Click **Pythonwebapp-env-1** in the list as it supports Python web applications. It could be different in your environment, based on the naming conventions:

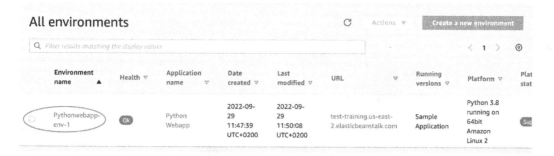

Figure 6.14 – Python All environments

4. Click the **Upload and deploy** button in order to follow the deployment process:

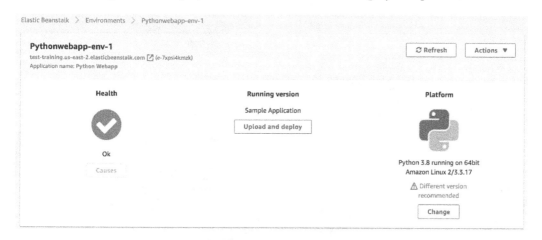

Figure 6.15 – Python web environment

5. In the **Upload and deploy** window, click the **Choose file** button:

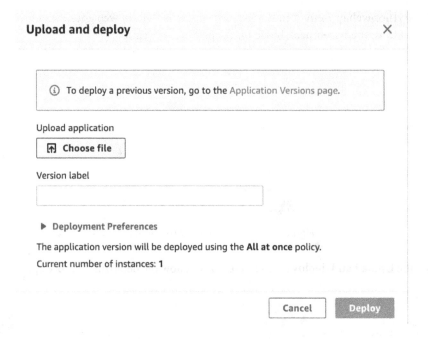

Figure 6.16 – Deploy environment

Once you click the **Choose file** button, your Python web application will be deployed to Elastic Beanstalk. As you can see in the following screenshot, you are going to select the local folder:

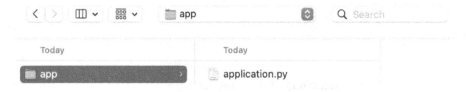

Figure 6.17 – Local folder

You can deploy whichever Python web framework you prefer, such as Flask, Django, and so on.

In this section, we learned how to deploy a custom Python web application to Elastic Beanstalk.

Summary

In this chapter, we learned about the AWS Elastic Beanstalk service and how to create a Python web environment in the cloud. Elastic Beanstalk is useful when you deploy web applications in the cloud. It comes with scalability, logging, and monitoring advantages. In the following chapter, we will take a look at how to monitor our applications via CloudWatch.

Part 3: Useful AWS Services to Implement Python

In this part, you will deep-dive into other AWS services for Python programming, such as monitoring, creating an API, database operations, and NoSQL with DynamoDB.

This part has the following chapters:

7

Monitoring Applications via CloudWatch

In this chapter, we are going to learn about one of the important AWS services, CloudWatch. CloudWatch is a serverless service that allows you to collect and monitor application logs within AWS. It has extensive integrations with most AWS services. When you start using any AWS service, it helps to observe an application via CloudWatch tools.

In this chapter, we are going to cover the following topics:

- What is CloudWatch?
- Collecting Lambda Logs via CloudWatch
- CloudWatch logs Insights
- CloudWatch alarms

What is CloudWatch?

When you deploy any application, it is important to track that it meets the set expectations regarding availability, performance, and stability. It is possible an issue may have occurred in the application. It's important to note that some of the AWS services could be down or run incorrectly. This is a very bad experience from a customer's point of view, and it would be better to observe these issues before the customer finds out. If you service an application via AWS, you need to use CloudWatch to monitor your applications to observe how they behave.

CloudWatch is a monitoring service in AWS; it provides different features to observe an application. The features of CloudWatch are as follows:

- Collecting and storing logs from AWS services such as Lambda and EC2.
- Providing a dashboard to monitor metrics and logs.

- The ability to create an alarm. For example, if an application has consumed significant memory on a server, you can create an alarm in order to be notified.

- The ability to correlate different metrics. For example, you can aggregate EC2 memory logs and CPU logs to have a better overall view of a situation.

- The detection of anomalous behavior with the machine learning-based CloudWatch anomaly detection feature.

Collecting Lambda logs via CloudWatch

In this topic, we are going to deploy a simple Python function in order to investigate logs via the CloudWatch service. Let's do so step by step:

1. Create a Lambda function in AWS. In *Chapter 3*, where we covered Lambda, the basic steps of the Lambda deployment were explained. Hence, here, we will provide a summary of the Lambda steps. The name of the Lambda function is `TestLogs`:

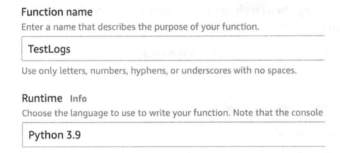

Figure 7.1 – Creating a Lambda function

2. The Lambda function creates a basic template, like the following:

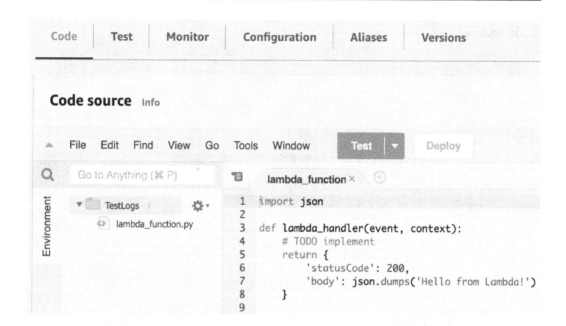

Figure 7.2 – The Lambda template

3. Copy the following code block to the handler:

```python
import json
import os

def lambda_handler(event, context):
    print('ENVIRONMENT VARIABLES')
    print(os.environ)

    return {
        'statusCode': 200,
        'body': json.dumps('Hello from Lambda!')
    }
```

os will import the operating system module; hence, you can see the environment variables via the logging print (os.environ) variable. Once we add the code block, Lambda code should be seen as follows:

Figure 7.3 – Lambda with logs

4.   Next, click the **Deploy** button to deploy the latest changes to Lambda and click the **Test** button. After testing the Lambda function, you are able to see the execution results:

Figure 7.4 – The execution results

Let's use the CloudWatch service to investigate the logs:

1. Open the CloudWatch service from AWS Management Console:

Figure 7.5 – The CloudWatch service

2. Click **Log groups** under the **Logs** dropdown in the left pane:

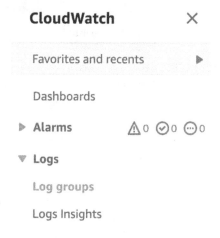

Figure 7.6 – The CloudWatch log group

3. Once you click **Log groups**, you will see a list. This list represents the running AWS services that create a log. In this list, find the Lambda function that you run:

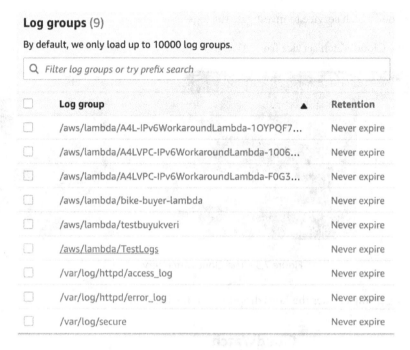

Figure 7.7 – Log list

4. Click **/aws/lambda/TestLogs**. The new page consists of the logs that Lambda creates. You
    can see a log stream. When the Lambda function runs, the logs are created in this list. At the
    beginning of the list, you can see the most up-to-date logs:

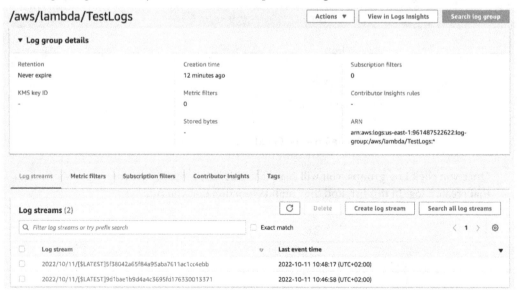

Figure 7.8 – The log page for Lambda

Let's click the latest link under **Log stream**:

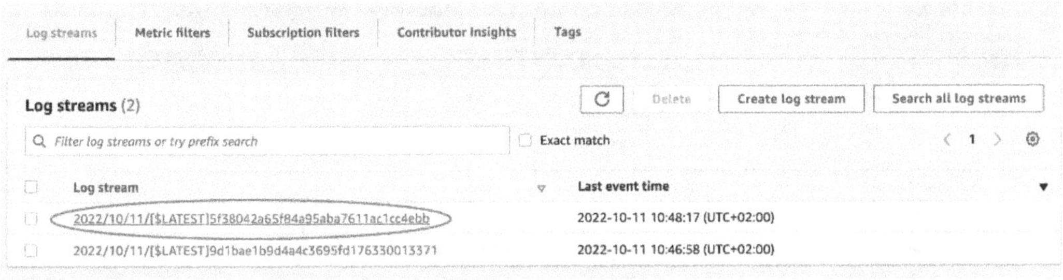

Figure 7.9 – Log stream

After clicking the link, you can see the detailed logs that Lambda creates:

Figure 7.10 – Lambda logs

This list shows a summary view of the log. When you click the down arrow to the left, the panel will open and you can investigate the detailed logs. In Lambda, we have logged the operating system variables for Lambda. Hence, you will see some details for that, such as region, memory size, and language:

Figure 7.11 – Log details

Congratulations! You are able to investigate Lambda logs via the CloudWatch service. It is simple to use CloudWatch to investigate a log for any AWS service. In the next topic, we will learn some tricks regarding filtering logs.

CloudWatch Log Insights

In this topic, we will take a look at **Log Insights**. If you have massive lines of logs, it is not easy to search and find the respective log that you are searching for. For this use case, Log Insights comes into play. CloudWatch Log Insights allows you to search logs with the filtering feature. Let's see how Log Insights helps us to search logs:

1.   Click **Log Insights** under the **Logs** dropdown in the left pane:

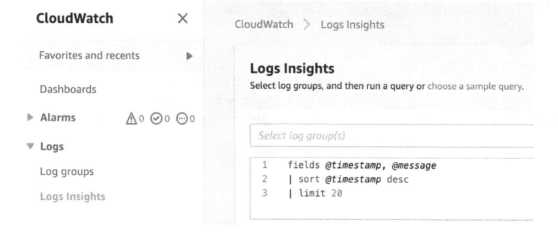

Figure 7.12 – Log Insights

2. Select the log that you want to investigate. In the previous example, we ran the TestLogs Lambda function, and I am also selecting that one here:

Logs Insights

Select log groups, and then run a query or choose a sample query.

5m	30m	1h	3h	12h	Custom

Select log group(s)   ▲

Q Type to search

☐ /aws/lambda/A4L-IPv6WorkaroundLambda-1OYPQF7PSZW6T

☐ /aws/lambda/A4LVPC-IPv6WorkaroundLambda-1006JHL3OUJIA

☐ /aws/lambda/A4LVPC-IPv6WorkaroundLambda-F0G3U13ZZ1IP

☐ /aws/lambda/bike-buyer-lambda

☐ /aws/lambda/testbuyukveri

☑ /aws/lambda/TestLogs

☐ /var/log/httpd/acce /aws/lambda/TestLogs

☐ /var/log/httpd/error_log

☐ /var/log/secure

All log groups loaded.

Figure 7.13 – The Log Insights window

3. Once you select it, you can see the default query:

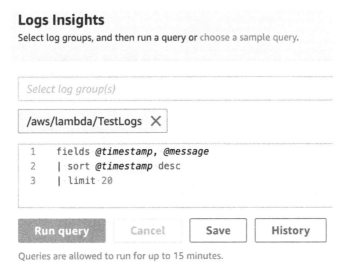

Figure 7.14 – The Log Insights filter

4. Click the **Run query** button in order to see the result. In this filter, `fields` represents the columns that will be listed, whereas the `sort` keyword indicates the sorting method, and you can see only 20 records with the `limit` keyword:

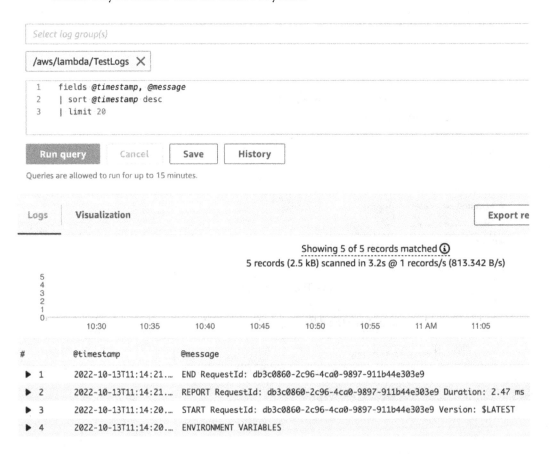

Figure 7.15 – Logs

Let's add one more filter to search for a keyword within the message. You can use the following query format:

```
fields @timestamp, @message
| filter @message like /AWS_DEFAULT_REGION/
| sort @timestamp desc
| limit 20
```

With this query, we search for logs that contain AWS_DEFAULT_REGION. Paste that and click **Run query** again. After running the query, you will see that the message lines are reduced:

Figure 7.16 – Filtered logs

When you expand the message, you will find what you searched for – in this case, AWS_DEFAULT_ REGION:

#	@timestamp	@message
▼ 1	2022-10-13T11:14:20.…	environ({'AWS_LAMBDA_FUNCTION_VERSION': '$LAT
	Field	Value
	@ingestionTime	1665652465080
	@log	961487522622:/aws/lambda/TestLogs
	@logStream	2022/10/13/[$LATEST]7e5aaadeaaf84a1d82aa2dbe
	@message	environ({'AWS_LAMBDA_FUNCTION_VERSION': '$LA
	@timestamp	1665652460997
	@xrayTraceId	1-6347d6ec-3d2886861f5141db1d518618
	_AWS_XRAY_DAEMON_ADDRESS	169.254.79.129
	_AWS_XRAY_DAEMON_PORT	2000
	_HANDLER	lambda_function.lambda_handler
	_X_AMZN_TRACE_ID	Root=1-6347d6ec-3d2886861f5141db1d518618;Par
	AWS_ACCESS_KEY_ID	ASIA57XI7JM7DMWDYZ50
	AWS_DEFAULT_REGION	us-east-1
	AWS_EXECUTION_ENV	AWS_Lambda_python3.9
	AWS_LAMBDA_FUNCTION_MEMORY_SIZE	128

Figure 7.17 – Detailed logs

As you can see, Log Insights is very helpful to search and filter logs within a massive log block. In the next topic, we will take a look at how to create an alarm.

CloudWatch alarms

AWS has more than 100 services, and it is not easy to control the behavior of all the services. You need to be informed if some AWS services achieve a specific metric. In *Chapter 4*, we covered how to create a server with an EC2 service. For example, you define a server for an EC2 service, and sometimes, its CPU usage is more than 90%, causing some performance problems. Another example would be to add a notification if you exceed a specific cost in AWS. For these kinds of scenarios, you can define a metric, and if the metric is reached, you will be notified via email.

In this topic, we are going to create an alarm to notify us if AWS cost exceeds $10 in a month. Let's implement the application:

1. Click **In alarm** under the **Alarms** dropdown in the **CloudWatch** pane:

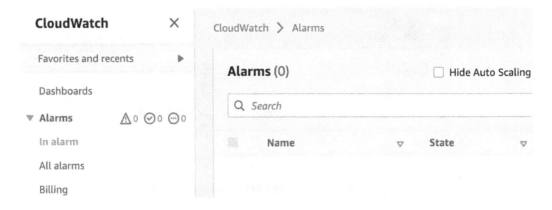

Figure 7.18 – In alarm

2. Click **Create Alarm**. You can click either the button to the right or the one at the bottom:

Figure 7.19 – Creating an alarm

3. Click the **Select metric** button:

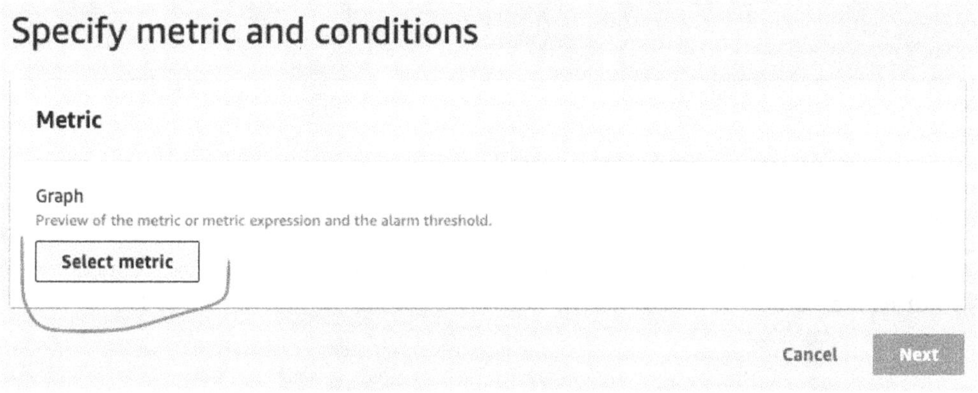

Figure 7.20 – Select metric

4. Once you click the **Select metric** button, you will be able to see a list of categories with which to narrow down your metric:

ApplicationELB	50	Billing	27
EC2	76	ElasticBeanstalk	2

Figure 7.21 – Metric types

In this list, you can see different types of metrics. **Billing** allows you to define cost-related metrics, while **Lambda** allows you to define Lambda-related metrics. In this example, we are going to define a monthly budget for our AWS account. The aim is to receive an alarm if our monthly cost exceeds a specific threshold:

1. Click **Billing** from the categories:

Billing	27
ElasticBeanstalk	2

Figure 7.22 – The Billing category

2. Click **Total Estimated Charge**. The intention is to define a metric if your total monthly AWS cost exceeds a target budget:

Figure 7.23 – Total Estimated Change

3. From the list, select **USD** and click **Select metric**. The currency type may vary, depending on your AWS account:

Figure 7.24 – The currency type

On the next screen, go to the **Define the threshold value** field. For this example, I added **10**, which means that if the total cost is greater than $10 for a month, an alarm will be activated. In this panel, you can also change the currency type, calculation type, and so on. In this case, the most important value is defining the target budget to receive an alarm. After you have done that, click the **Next** button:

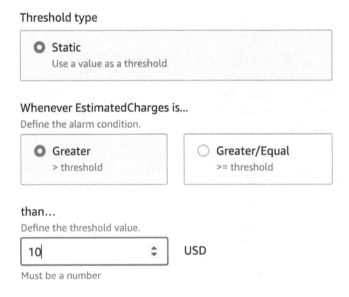

Figure 7.25 – Threshold value

4. In the next panel, we are going to define the alarm endpoint. In this case, we have selected the **Create new topic** radio button. **Simple Notification Service (SNS)** is used to communicate between services and end users. This is a choice under **Send a notification to the following SNS**. Once we select **Create new topic**, we can define an email address in the **Email endpoints that will receive the notification…** section. SNS is an access point to filter messages in order to send them to different subscribers such as Lambda or email. You can keep the topic name as is; it is the same as the SNS topic name. When completed, click **Create topic**:

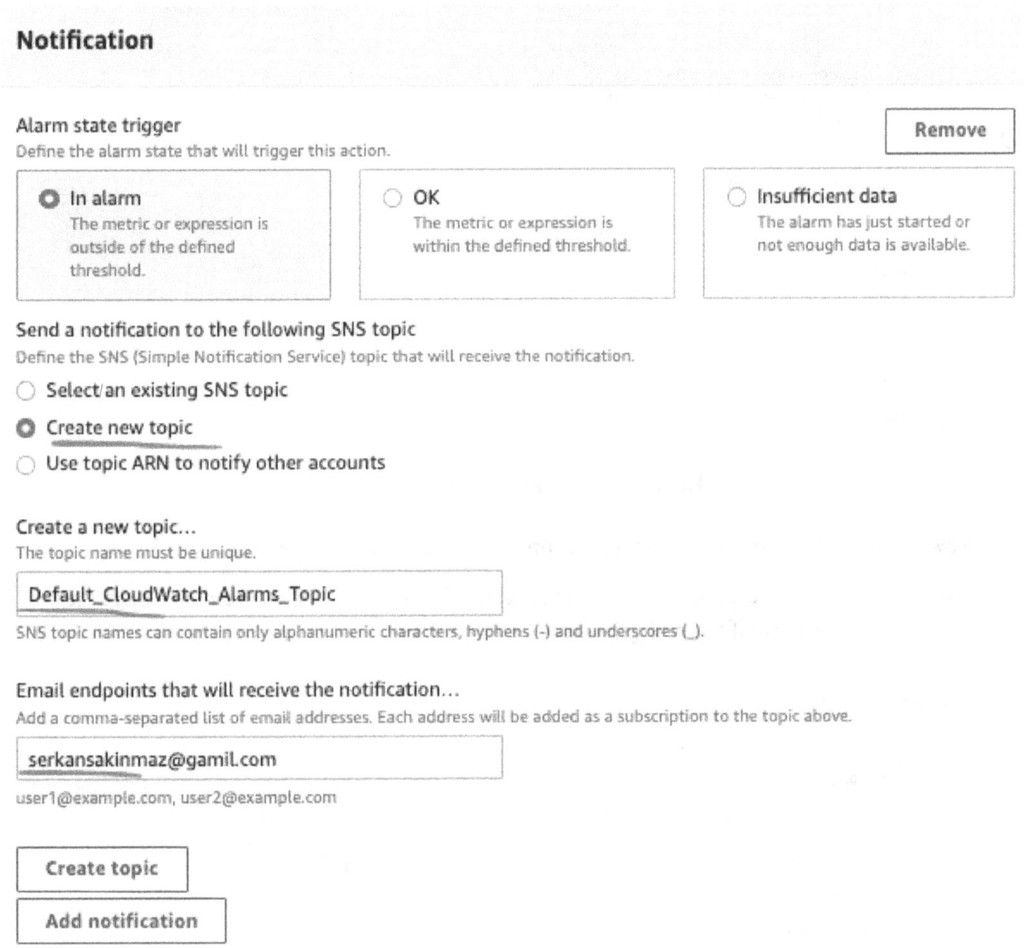

Notification

Alarm state trigger
Define the alarm state that will trigger this action.

Remove

- ● **In alarm**
  The metric or expression is outside of the defined threshold.

- ○ **OK**
  The metric or expression is within the defined threshold.

- ○ **Insufficient data**
  The alarm has just started or not enough data is available.

Send a notification to the following SNS topic
Define the SNS (Simple Notification Service) topic that will receive the notification.

○ Select an existing SNS topic

● Create new topic

○ Use topic ARN to notify other accounts

Create a new topic...
The topic name must be unique.

Default_CloudWatch_Alarms_Topic

SNS topic names can contain only alphanumeric characters, hyphens (-) and underscores (_).

Email endpoints that will receive the notification...
Add a comma-separated list of email addresses. Each address will be added as a subscription to the topic above.

serkansakinmaz@gamil.com

user1@example.com, user2@example.com

Create topic

Add notification

Figure 7.26 – Receiver

5.   After **Create topic** is clicked, AWS will create an endpoint in order to send an email:

Figure 7.27 – Creating an endpoint

Now, you have an endpoint, and you can proceed by clicking the **Next** button.

6.   The next step is to define the alarm name. In this case, I named it `BillingAlarmGreaterThan10`, since it sends an alarm if the billing cost goes above than $10:

Figure 7.28 – Naming the alarm

7. The next step is to review the input and click **Create alarm**:

Figure 7.29 – Creating the alarm

8. If you successfully create the alarm, you will be redirected to the **Alarm** list to see the alarm that you created. We can see the alarm as follows:

Figure 7.30 – The billing alarm type

In this topic, we have created an alarm. An alarm is useful if we need to create a notification for the AWS service behaviors. This example will send a notification if, for example, we reach the defined cost limit.

Summary

In this chapter, we learned about the AWS CloudWatch service and how to investigate service logs in AWS. CloudWatch is very useful for logging; it also allows you to define some metrics and alarms to monitor services. In the following chapter, we will take a look at database operations within AWS.

Database Operations with RDS

In this chapter, we are going to learn the basics of **Amazon Relational Database Service** (**Amazon RDS**) and create an RDS instance in order to make a database operation. You can use RDS to create the most popular databases in AWS. You can create Oracle, MySQL, or MS SQL databases on the cloud with scaling capabilities. In general, when you need to create a database, you must manage the infrastructure using an on-premises system. Managing the hardware and infrastructure, installing the database, and then monitoring could require a lot of effort to set up. AWS allows you to select the database type that you want and then create it with a simple button click – that is all:

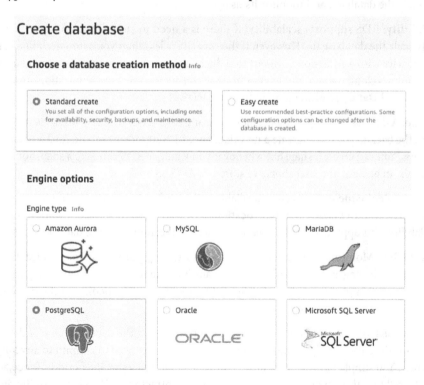

Figure 8.1 – Click to create

In this chapter, we will create a database and make some operations using Python. The chapter covers the following topics:

- Features of RDS
- Provisioning RDS
- Connecting to the RDS
- Creating a table in the database
- Database operations with Python
- Secrets Manager

Features of RDS

RDS comes with different features that facilitate the creation and maintenance of the database. Let's look at the most important features:

- **Easy to use**: You can easily create and maintain RDS via the AWS console. It also allows us to use some API capabilities to make some programmatic operations. For example, you can create and scale the database, and monitor its usage.

- **Scalability**: RDS supports scalability; if there is a need to support more capacity, you can easily scale the database up. However, if the capacity is less than you estimate, you can reduce the capacity with a *scale-down request* to reduce the cost. Another option is Amazon Aurora, which allows cloud users to implement more performance-intensive applications that support a **Relational Database Management System (RDBMS)**.

- **Backup**: A database backup is important in case any issue arises with the infrastructure. In some cases, the backup is used to create a new database. RDS supports both manual and automated backups. You can create a snapshot whenever you want, or RDS can take a snapshot at regular intervals. In general, the snapshots are stored in AWS S3 buckets.

- **Multi-AZ deployment**: RDS can be available within different locations to improve availability. If the infrastructure is down in one location, RDS can serve in another location to improve availability. This approach can be used for critical applications that use databases in the cloud.

- **Monitoring**: Monitoring is very important for critical applications. You can track how the database is behaving and see whether there are any issues in it. RDS has a supporting monitoring feature. For example, you can track when I/O problems are happening in the database, and you can take the right action.

- **Cost options**: AWS offers different pricing options for using the database. One of the popular options is the *pay-as-you-go* option. In this option, you don't need to commit to any long-term contract. You simply pay for how many resources you use in a specific period. Hence, you can pay the bill monthly. In other options, you make a contract with AWS for a specific duration; however, in this case, you have to pay for the contract even if you don't use the database.

Provisioning RDS

In this section, we are going to create a sample relational database on the cloud. To provision the RDS on AWS, carry out the following steps:

1. Open the AWS console and type `rds` in the search box:

Figure 8.2 – RDS on the console

2. Click **Databases** on the left pane to see the list of databases. To create a new database, click **Create database**:

Figure 8.3 – Database list

3. On the new panel, **Create database**, and fill out the information required for the new database. RDS supports multiple database types, such as Amazon Aurora, MySQL, MariaDB, PostgreSQL, Oracle, and Microsoft SQL Server. In this example, we will use **MySQL**:

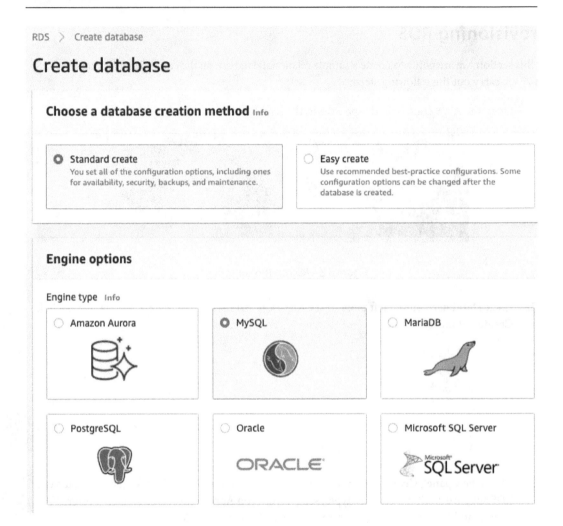

Figure 8.4 – MySQL selection

4.   After selecting **MySQL**, scroll down and select the correct version of MySQL. In this example, we will use one of the latest versions, **MySQL 8.0.28**:

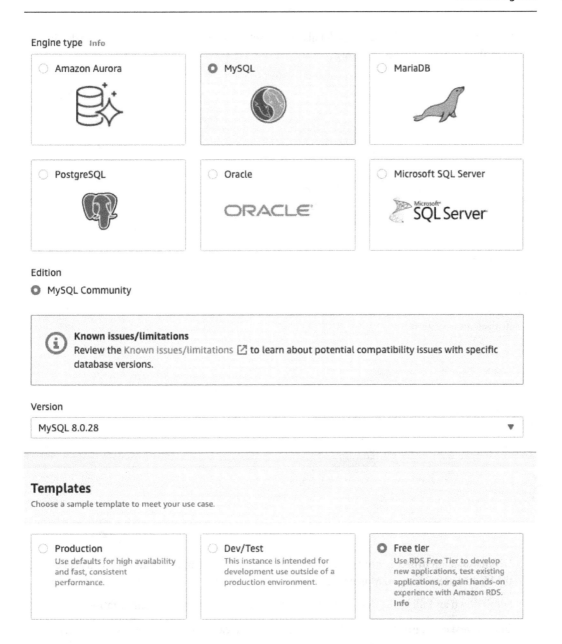

Figure 8.5 – Template selection

5. Templates are useful for working in different environments. When you select the **Production** template, it gives you high availability. In this example, we will select the **Free tier** template to avoid any costs.

6.   Scroll down and fill in the **Settings** details. In the **Settings** panel, you need to fill in the database identifier, username, and password:

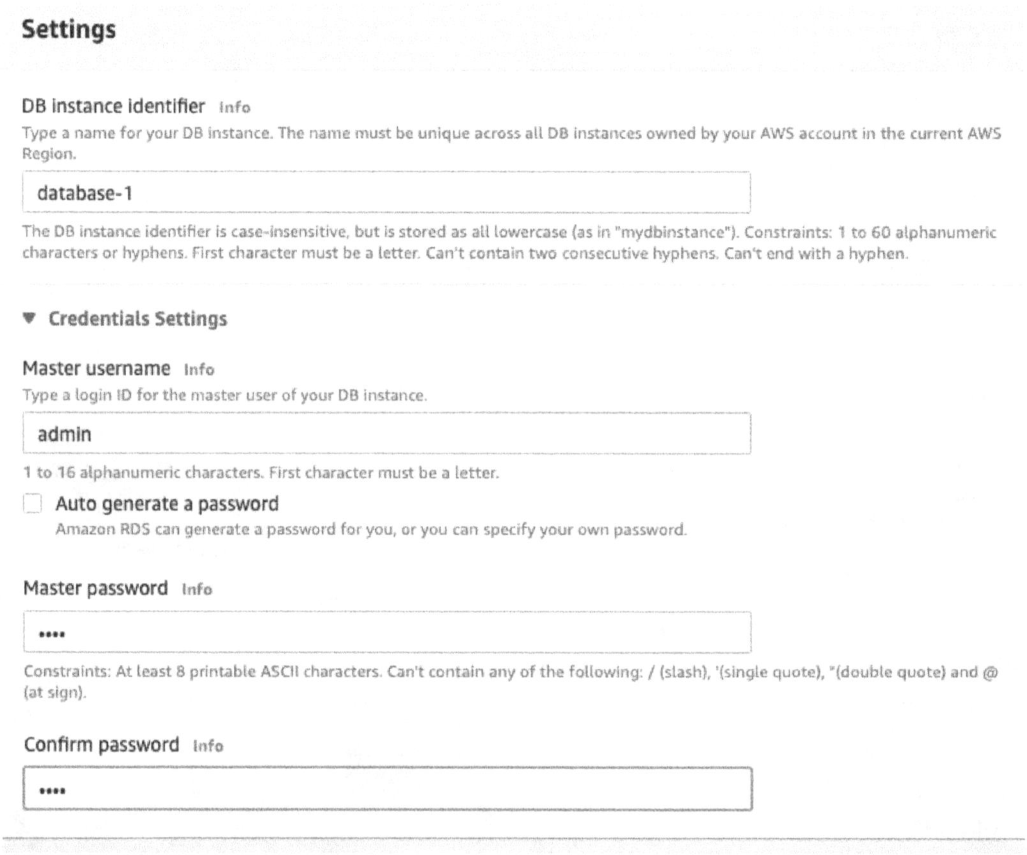

Settings

DB instance identifier Info

Type a name for your DB instance. The name must be unique across all DB instances owned by your AWS account in the current AWS Region.

> database-1

The DB instance identifier is case-insensitive, but is stored as all lowercase (as in "mydbinstance"). Constraints: 1 to 60 alphanumeric characters or hyphens. First character must be a letter. Can't contain two consecutive hyphens. Can't end with a hyphen.

▼ **Credentials Settings**

Master username Info

Type a login ID for the master user of your DB instance.

> admin

1 to 16 alphanumeric characters. First character must be a letter.

☐ **Auto generate a password**

Amazon RDS can generate a password for you, or you can specify your own password.

Master password Info

> ••••

Constraints: At least 8 printable ASCII characters. Can't contain any of the following: / (slash), '(single quote), "(double quote) and @ (at sign).

Confirm password Info

> ••••

Figure 8.6 – Settings

DB instance identifier is used to represent the database name in the cloud. You can also enter the **Master username** and **Master password** details. These credentials are important for security.

Scroll down and fill in the details with regard to storage and instance configuration.

7.   In **Instance configuration**, in **DB instance class**, you can select the processor and memory types. Since we are creating it for education, you can select the simple instance type that has basic hardware features. Another hardware selection is made for **Storage**. You can keep what AWS has recommended or use the minimum values.

In **Storage type**, you can select the disk type. In **Allocated storage**, you have to specify the limit of the disk. For this example, we selected **200** GiB. If the disk needs to be scaled, you can check the **Enable storage autoscaling** checkbox.

When the disk is scaled, the value you enter in the **Maximum storage threshold** field is the maximum value of the database. In this case, the maximum threshold is 1000:

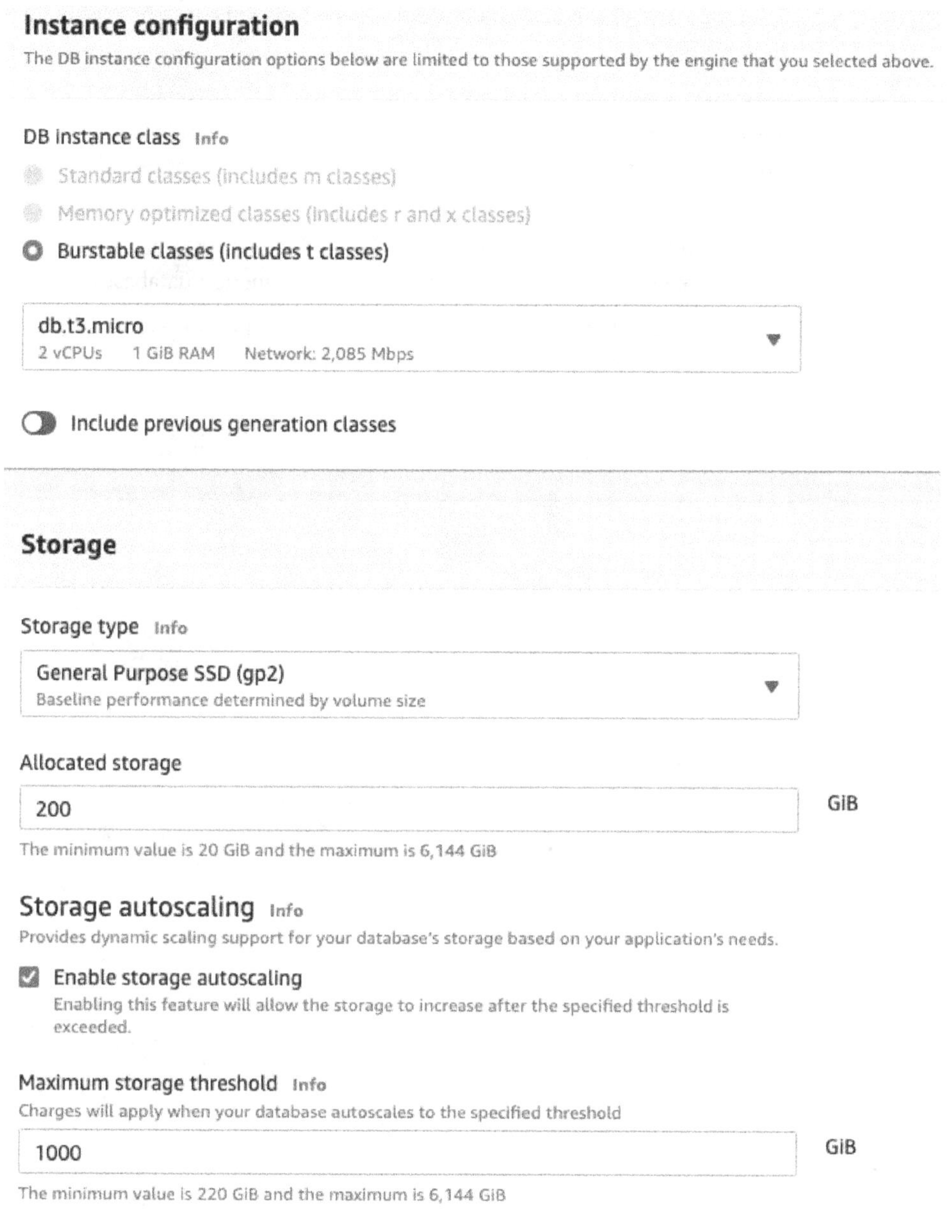

Instance configuration

The DB instance configuration options below are limited to those supported by the engine that you selected above.

DB instance class Info

○ Standard classes (includes m classes)

○ Memory optimized classes (includes r and x classes)

⦿ Burstable classes (includes t classes)

| db.t3.micro | ▼ |
| 2 vCPUs   1 GiB RAM   Network: 2,085 Mbps | |

◑ Include previous generation classes

Storage

Storage type Info

| General Purpose SSD (gp2) | ▼ |
| Baseline performance determined by volume size | |

Allocated storage

| 200 | GiB |

The minimum value is 20 GiB and the maximum is 6,144 GiB

Storage autoscaling Info
Provides dynamic scaling support for your database's storage based on your application's needs.

☑ Enable storage autoscaling
  Enabling this feature will allow the storage to increase after the specified threshold is exceeded.

Maximum storage threshold Info
Charges will apply when your database autoscales to the specified threshold

| 1000 | GiB |

The minimum value is 220 GiB and the maximum is 6,144 GiB

Figure 8.7 – Instance configuration (part 1)

Scroll down and fill in the details with regard to **Connectivity**.

8. In the first option, AWS asks whether you want to connect to EC2. For this example, we don't need to connect to EC2, so we select **Don't connect to an EC2 compute resource**. (After setting up the database, we will use Lambda for database operations.) RDS needs to be created in the VPC, so in **Virtual private cloud (VPC)**, we select **Create new VPC**, and it will automatically create a VPC.

    Another option is to select a group in **DB Subnet group**. This allows you to define which IP group is going to connect to the database. It is also important in terms of security. You can limit the IP range with this option.

    Public access allows you to enable access over the internet. For this application, we will use public access. However, you need to be careful when you set production databases as public.

    The final option for **Connectivity** is to select a group in **VPC security group (firewall)**. In this case, you can define the same security group that connects to RDS:

Connectivity Info

Compute resource

Choose whether to set up a connection to a compute resource for this database. Setting up a connection will automatically change connectivity settings so that the compute resource can connect to this database.

- ● **Don't connect to an EC2 compute resource**
  Don't set up a connection to a compute resource for this database. You can manually set up a connection to a compute resource later.

- ○ **Connect to an EC2 compute resource**
  Set up a connection to an EC2 compute resource for this database.

Virtual private cloud (VPC) Info

Choose the VPC. The VPC defines the virtual networking environment for this DB instance.

| Create new VPC ▼ |

Only VPCs with a corresponding DB subnet group are listed.

> ⓘ After a database is created, you can't change its VPC.

DB Subnet group Info

Choose the DB subnet group. The DB subnet group defines which subnets and IP ranges the DB instance can use in the VPC that you selected.

| Create new DB Subnet Group ▼ |

Public access Info

- ● **Yes**
  RDS assigns a public IP address to the database. Amazon EC2 instances and other resources outside of the VPC can connect to your database. Resources inside the VPC can also connect to the database. Choose one or more VPC security groups that specify which resources can connect to the database.

- ○ **No**
  RDS doesn't assign a public IP address to the database. Only Amazon EC2 instances and other resources inside the VPC can connect to your database. Choose one or more VPC security groups that specify which resources can connect to the database.

VPC security group (firewall) Info

Choose one or more VPC security groups to allow access to your database. Make sure that the security group rules allow the appropriate incoming traffic.

- ● **Choose existing**
  Choose existing VPC security groups

- ○ **Create new**
  Create new VPC security group

Existing VPC security groups

| Choose one or more options ▼ |

| default ✕ |

Figure 8.8 – Instance configuration (part 2)

Scroll down and fill in the database port information.

9. **Database port** defines which port is used to connect to the database. The default value is 3306 for MySQL, but you can also change it:

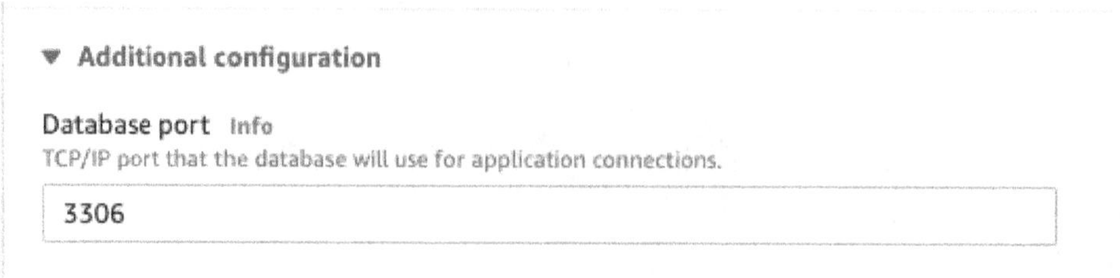

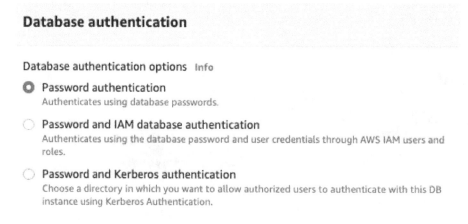

Figure 8.9 – Database port

Scroll down and fill in the authentication details.

10. **Database authentication** is used to define the approach for password management. You can connect with only a password, a combination of a password with IAM authentication, or a password with Kerberos authentication. Let's keep it simple and just use **Password authentication**:

Database authentication

Database authentication options   Info

○ Password authentication
Authenticates using database passwords.

○ Password and IAM database authentication
Authenticates using the database password and user credentials through AWS IAM users and roles.

○ Password and Kerberos authentication
Choose a directory in which you want to allow authorized users to authenticate with this DB instance using Kerberos Authentication.

Figure 8.10 – Database authentication

Scroll down and fill in the details regarding database creation.

11. As a final step, you can keep other values as is. Click **Create database** and proceed with the database creation:

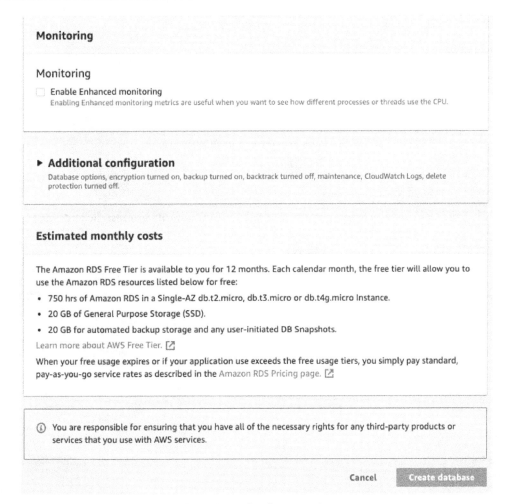

Figure 8.11 – Database creation

This forwards you to the **Databases** list, in which you can see the database is being created:

Figure 8.12 – Databases list with a Creating status

After some time, you can see the database is ready to use:

Figure 8.13 – Databases list with an Available status

We will connect from our computer. To connect to the database, we need to enable the connection from outside of AWS.

12. Click the **Connectivity & security** tab. You will see **VPC security groups**; click the link:

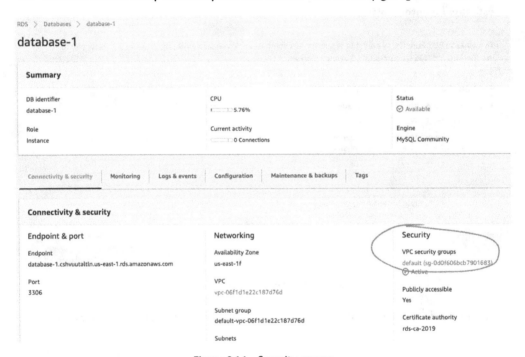

Figure 8.14 – Security groups

13. In the new panel, click **Edit inbound rules**. This will allow us to define the inbound connections:

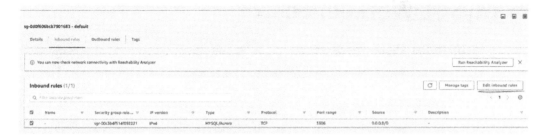

Figure 8.15 – Inbound rules

14. Add the rule for the MySQL/Aurora type and click **Save**, which isn't depicted in the following figure but is situated at the bottom of the page:

Figure 8.16 – Adding the rule

These steps allow us to accept the connection from outside of AWS. Hence, we will connect to AWS via a local computer.

Congrats! You have created the database on the cloud. As you can see in the steps, creating a database is easy and efficient on the cloud. Let's connect to the database in the next topic.

Connecting to the RDS

In this section, we are going to connect to the RDS from one of the database viewers. For that purpose, you can install a free database viewer; I will use a MySQL viewer. To install the MySQL viewer, carry out the following steps:

1. Open the following link: `https://www.mysql.com/products/Workbench/`.

2. Click **Download Now** on the main page:

Figure 8.17 – MySQL Workbench

3. Click **Download** on the next page:

Figure 8.18 – MySQL Workbench download

4. Double-click and install the downloaded package, and the installation will be done.

5. Once the installation has been completed, click the + symbol to connect to the new database:

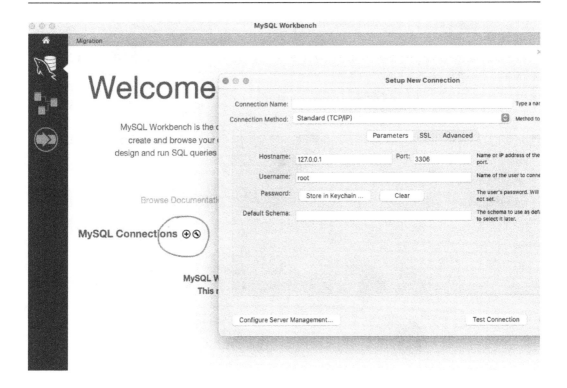

Figure 8.19 – New connection

6.   Open AWS and copy the connection details:

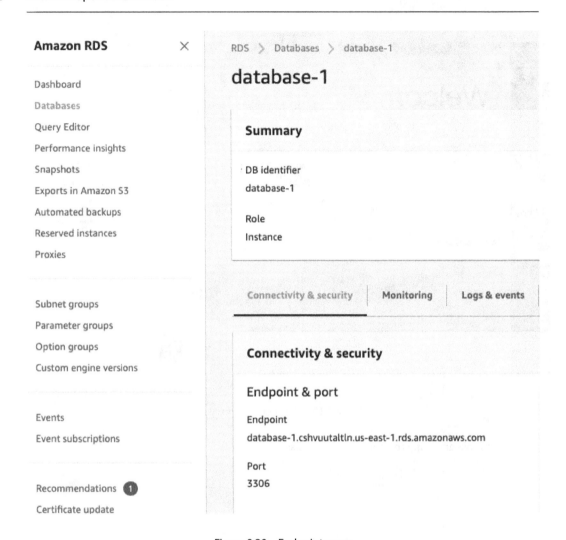

Figure 8.20 – Endpoint name

7.   Fill out the endpoint and password details in MySQL Workbench and click **Test Connection**:

Setup New Connection

Connection Name: database-1   Type a name for the connection

Connection Method: Standard (TCP/IP)   ⬍   Method to use to connect to the RDBMS

Parameters   SSL   Advanced

Hostname: database-1.cshvuutaltln.us-east   Port: 3306   Name or IP address of the server host - and TCP/IP port.

Username: admin   Name of the user to connect with.

Password: [Store in Keychain ...] [Clear]   The user's password. Will be requested later if it's not set.

Default Schema: [   ]   The schema to use as default schema. Leave blank to select it later.

[Configure Server Management...]   [Test Connection]   [Cancel]   [OK]

Figure 8.21 – Test Connection

After clicking **Test Connection**, you will be able to see the connection:

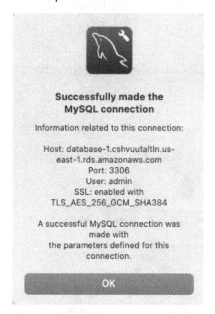

**Successfully made the
MySQL connection**

Information related to this connection:

Host: database-1.cshvuutaltln.us-
east-1.rds.amazonaws.com
Port: 3306
User: admin
SSL: enabled with
TLS_AES_256_GCM_SHA384

A successful MySQL connection was
made with
the parameters defined for this
connection.

OK

Figure 8.22 – Connection is successful

Good work! We have successfully connected to the RDS database from MySQL Workbench. Let's create a table and insert some records in the next topic.

Creating a table in the database

We have created a database in the cloud and have connected via MySQL Workbench. As a next step, we are going to create a table via MySQL Workbench:

1. Connect to the database via MySQL Workbench.

2. Create a database with the following command and click the *lightning* symbol, as shown in the figure that follows:

```
CREATE DATABASE address;
```

Figure 8.23 – Creating a database

3. Execute the USE address command in order to switch databases:

```
USE address;
```

Figure 8.24 – USE address

4. Create an address table:

```
CREATE TABLE address (id INT, address VARCHAR(20));
```

Figure 8.25 – Creating a table

We have created an address table, and for the next step, we are going to insert data into the table.

5.   Execute the following script to insert data into the table:

```
INSERT INTO address (id,address) VALUES(1,"Germany");
INSERT INTO address (id,address) VALUES(2,"USA");
```

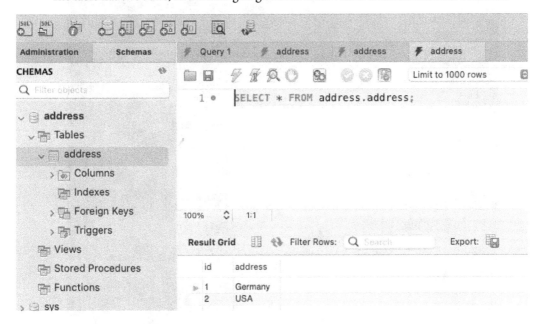

Figure 8.26 – Inserting script

The table has two rows, and we are going to read these values from the Lambda function:

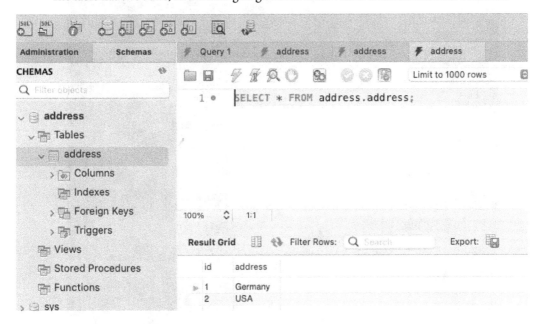

Figure 8.27 – Select script

In this topic, we have created a simple table and inserted records. The insertion was made with MySQL Workbench, but you can also use other database tools. As a next step, we are going to read the records using Python.

Database operations with Python

In this section, we are going to read a table using Python. To execute a Python function, we will use PyCharm on a local computer. Carry out the following steps:

1. Open PyCharm or an IDE, whichever you prefer.

2. We are going to install MySQL Connector to PyCharm. MySQL Connector will be used for database operations from Python. In PyCharm, select **File | New Projects Setup | Preferences for New Projects...**:

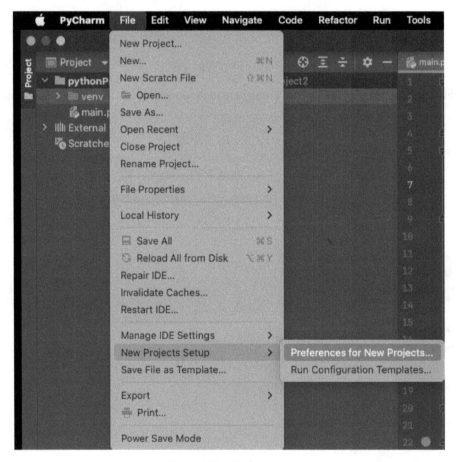

Figure 8.28 – Preferences

3. In the panel, select **Python Interpreter**:

Figure 8.29 – Python Interpreter

4. To add a new package, click the + symbol:

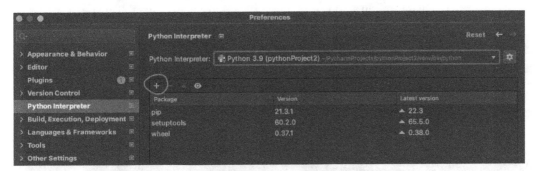

Figure 8.30 – Adding a package

5. In the upcoming panel, type mysql-conn to install **mysql-connector**. You will be able to see **mysql-connector**. Click **Install Package** to install it:

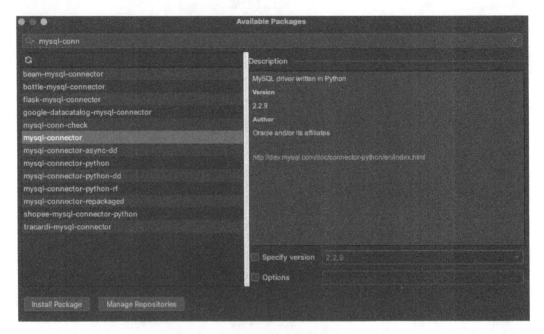

Figure 8.31 – Installing mysql-connector

6.   Once you install it, you will be able to see **mysql-connector** within the installed packages:

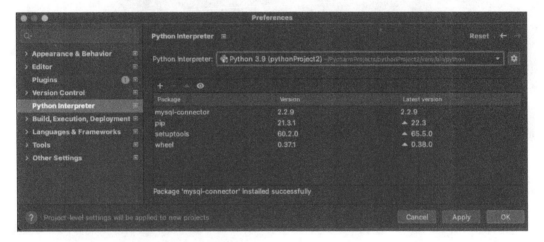

Figure 8.32 – Package list

7. Copy and paste the following code to read data from the database:

```
import mysql.connector

#rds settings
rds_host = "database-1.********.us-east-1.rds.amazonaws.com"
name = "**min"
password = "*****234"
db_name = "address"


if __name__ == '__main__':
    conn = mysql.connector.connect(host=rds_host, user=name,
passwd=password, database=db_name, port=3306)
    cursor = conn.cursor()
    cursor.execute("select * from address")
    data = cursor.fetchall()

    print(data)
```

The preceding code block connects to the RDS database and reads from the address table by executing the select * from address query. For rds_host, name, and password, please fill out your database host and credentials:

```
import mysql.connector


#rds settings
rds_host = "database-1.cshvuutaltln.us-east-1.rds.amazonaws.com"
name = "admin"
password = "Arif1234"
db_name = "address"


if __name__ == '__main__':
    conn = mysql.connector.connect(host=rds_host, user=name, passwd=password, database=db_name, port=3306)
    cursor = conn.cursor()
    cursor.execute("select * from address")
    data = cursor.fetchall()

    print(data)
```

Figure 8.33 – Query from the database

8.   When you click **Run**, you can see the results from the database:

Figure 8.34 – Results from the database

Congrats! You are able to read data from the AWS database via Python. You can also extend your query by implementing `insert` and `update` queries. In this topic, we learned how to make a database operation via Python.

Secrets Manager

Secrets Manager is an AWS service that allows you to manage and retrieve database credentials, which can be helpful when using a database. Let's learn how to use Secrets Manager:

1.   Open **Secrets Manager** via the console:

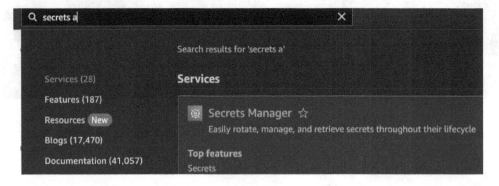

Figure 8.35 – Opening Secrets Manager

2. Click the **Store a new secret** button:

Description	Last retrieved (UTC)

No secrets

Store a new secret

Figure 8.36 – Storing a new secret

3. Select the secret type that you want to store a secret for, and fill out the username and password. In this case, we will select the **database-1** instance. After filling out the details, click **Next**:

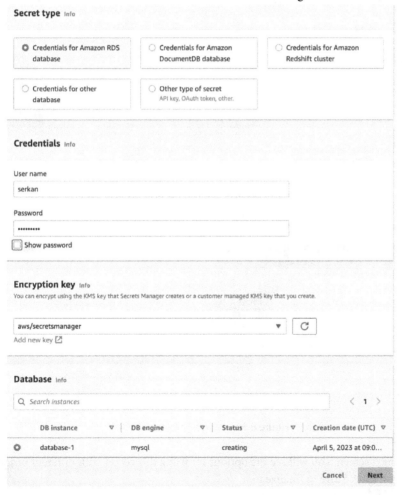

Figure 8.37 – Filling out the details

4. You need to give a name to the upcoming path in the **Secret name** textbox:

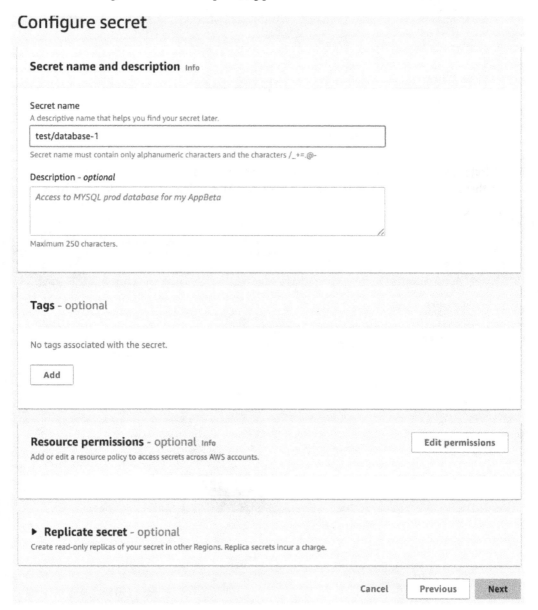

Figure 8.38 – Naming the secret

5. On the next screen, you will see the options for using this secret with different programming languages. Click **Store** to finalize it:

Figure 8.39 – Store secret

6. As the final step, you will see the secret on the list:

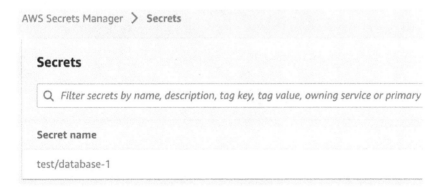

Figure 8.40 – List of secrets

Congrats! You have learned how to create and store secrets on the cloud in a secure way.

Summary

In this chapter, we learned about AWS RDS, which is used to create a relational database on the cloud. You can create your database in an efficient way. The point to note is that you have the possibility to create different databases, including MySQL, Microsoft SQL, and PostgreSQL. In this chapter, we have created an RDS instance on the cloud and run a Python application to make a read operation. In the following chapter, we will take a look at creating an API in AWS.

9
Creating an API in AWS

In this chapter, we are going to learn how to create an **application programming interface (API)** via **API Gateway**. API Gateway is an AWS service that allows you to create and maintain an API. With the API Gateway service, you don't need to provision a server; AWS manages it in the backend. In addition to that, API Gateway helps you to monitor incoming and outgoing requests. Another advantage of API Gateway is to scale up your API services when there is a huge request from users.

The chapter covers the following topics:

- What is API Gateway?
- Creating an API using API Gateway

What is API Gateway?

API Gateway is an AWS service that is used to create, maintain, and publish an API. API Gateway supports multiple API protocols, such as **RESTful** (also known as the REST API) and **WebSocket**.

API Gateway is a single point of entry for the backend services. As you can see in the following architecture, API Gateway gets a request from a client and integrates the incoming request with microservices, databases, AWS Lambda, or another AWS service:

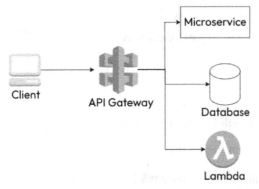

Figure 9.1 – Architecture of API Gateway

Now that we have a good idea of what API Gateway is, let's have a look at its features.

Features of API Gateway

The features of API Gateway are as follows:

- It supports different protocols, such as RESTful and WebSocket.

- You can monitor incoming and outgoing API requests, which enhances the visibility of the service.

- You can easily create and maintain the API. It can be created either in AWS Management Console or the AWS CLI.

- Security is important for cloud services, as well as the API. You can create a key to enable secure access to the API. In addition to that, you can add an SSL certificate to verify the request.

- It has built-in integration with AWS services. When you implement an API, you can easily integrate it with AWS services.

- It is a scalable service that adds more resources when you have more requests. For example, on Black Friday, there is more load on e-commerce websites. In these cases, API Gateway automatically scales your API requests. In this case, you can also define a **Cross-Origin Resource Sharing** (**CORS**) policy as a security feature that controls the HTTP request.

In this section, we have looked at the basic features of API Gateway, and now we will start to implement sample API applications.

Creating an API using API Gateway

We are going to create a simple API that accepts a request from a client. The API accepts two numbers, sums up two numbers in a Lambda function, and returns the calculated values. AWS Lambda is going to be implemented via Python. You can see the high-level flow in the following architecture:

Figure 9.2 – Data flow

We are going to start with the Lambda function creation. After the Lambda function creation, API Gateway is going to be set up with Lambda integration.

Let's create the Lambda function step by step:

1. Open the console and navigate to the **AWS Lambda** page:

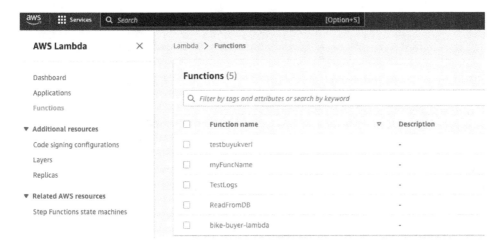

Figure 9.3 – Lambda function

2. Create a new Lambda function. Let's name it SumUpLambda:

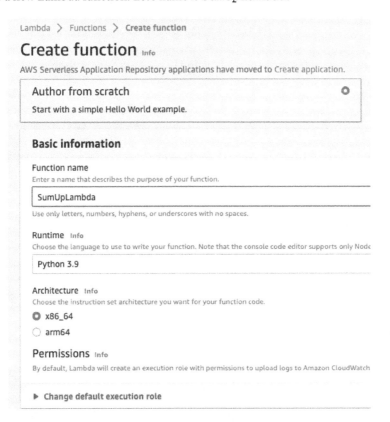

Figure 9.4 – Creating a new Lambda function

3.   Click **Create function** and wait a few seconds while the function is created:

Figure 9.5 – Clicking Create function

A few seconds later, you will see the Lambda function has been created with the template code:

Figure 9.6 – Lambda template

Let's create a Lambda function that sums up two values:

```python
import json

def lambda_handler(event, context):
    number1 = event['Number1']
    number2 = event['Number2']
    sum = number1 + number2

    return {
        'statusCode': 200,
        'Sum': sum
    }
```

This code snippet takes two numbers as parameters, such as Number1 and Number2. The Lambda function calculates the sum of two values and returns a status code and the value of the sum. When we call this function from the API, it returns the sum value as well as statusCode.

Let's paste this code block into the Lambda function:

Figure 9.7 – Actual Lambda code

Now, let's follow these steps:

1. Click **Test**. A new panel opens in which Lambda asks for a test parameter:

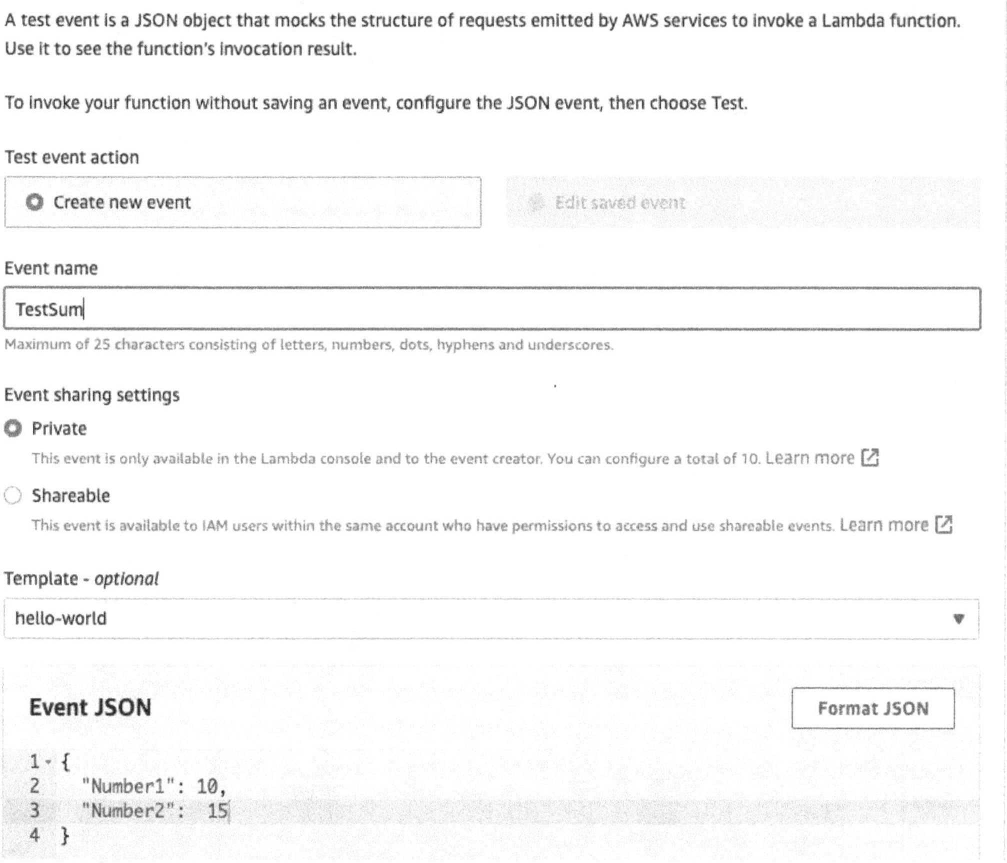

A test event is a JSON object that mocks the structure of requests emitted by AWS services to invoke a Lambda function. Use it to see the function's invocation result.

To invoke your function without saving an event, configure the JSON event, then choose Test.

Test event action

○ Create new event

⊙ Edit saved event

Event name

TestSum

Maximum of 25 characters consisting of letters, numbers, dots, hyphens and underscores.

Event sharing settings

⦿ Private

This event is only available in the Lambda console and to the event creator. You can configure a total of 10. Learn more ↗

○ Shareable

This event is available to IAM users within the same account who have permissions to access and use shareable events. Learn more ↗

Template - *optional*

hello-world ▼

Event JSON

Format JSON

```
1 ▾ {
2     "Number1": 10,
3     "Number2":  15
4 }
```

Figure 9.8 – Test event

2. As can be seen in the preceding figure, you can paste the following JSON to see whether the Lambda function is running properly before integrating with the API:

```
{
    "Number1": 10,
    "Number2": 15
}
```

3. Click **Save**, which is under the **Event JSON** panel:

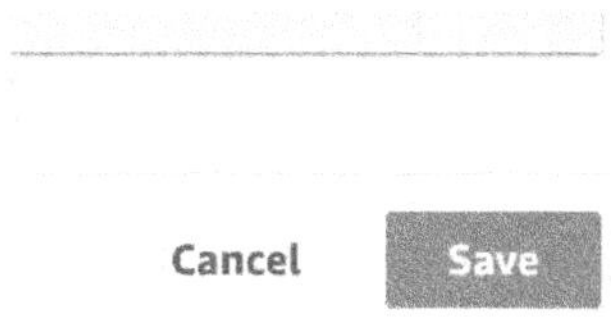

Cancel   Save

Figure 9.9 – Clicking on the Save button

4. Deploy the changes by clicking **Deploy**:

Figure 9.10 – Deploying Lambda

After the Lambda deployment, we are going to integrate API Gateway with AWS Lambda. Lambda will be used as the backend for API Gateway.

Let's create an API step by step:

1. Open the console and search for `api gateway`:

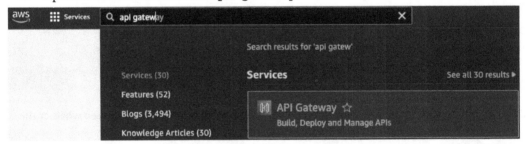

Figure 9.11 – The console

2. On the main screen, select **REST API**, and click **Build**:

Figure 9.12 – REST API

3. You will now see a new screen to be filled out. We will select **New API** in the **Create new API** section. Other options in this section allow you to create an example API or import a predefined API. In the **Settings** section, we will add the **API name** and **Description** details. In the **Endpoint Type** drop-down list, we will select **Regional**, which is used to create an API that is accessible from the same region:

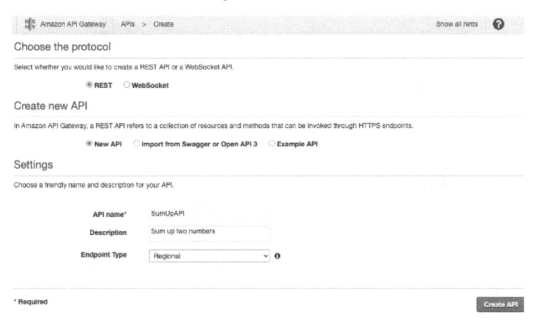

Figure 9.13 – Form for API creation

4. Once you click **Create API** (as depicted in the preceding figure), you will be taken to a new page that allows you to define the details for a custom SumUp API:

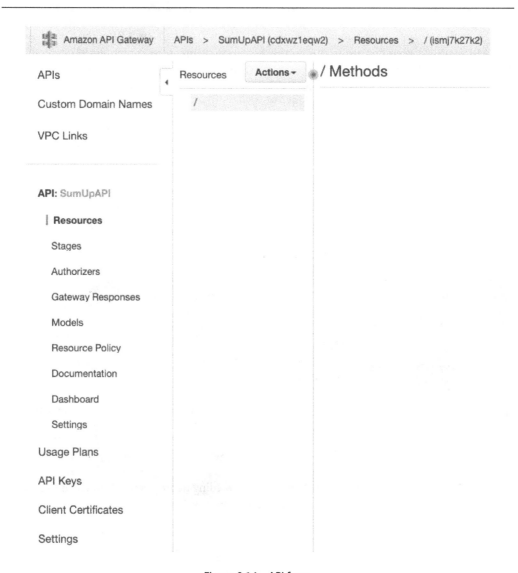

Figure 9.14 – API form

5.   Now, we are going to define the API details. Click on the **Actions** dropdown and select **Create Method**:

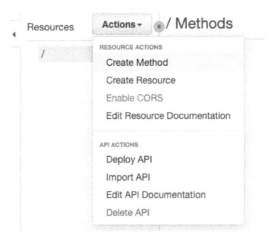

Figure 9.15 – Create Method

6.   When we create a method, we select **POST** as the API type:

Figure 9.16 – Selecting POST

While you implement an API, you can select API types. The following are the most used API types:

- **GET** is used to retrieve data from a source.

- **POST** is used to send data to a source. In our example, **POST** will bring the calculation of SumUp from Lambda.

- **PUT** is used to update the data in a source.

- **DELETE** is used to delete the data in a source.

7. When you select **POST**, you need to choose the integration type. For this example, we are going to select the **Lambda Function** integration type:

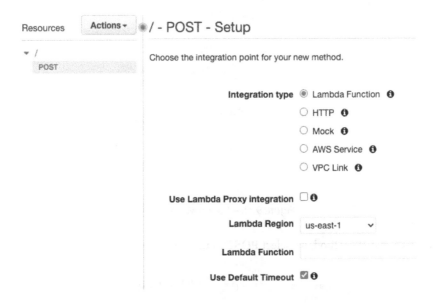

Figure 9.17 – Setting up the integration type

8. Select the **SumUpLambda** function that is implemented, and click **Save**, which is not depicted in the following figure but is situated at the bottom of the page:

Figure 9.18 – Selecting Lambda

9. When you click **Save**, it asks for confirmation to allow the required permissions. Click **OK** and it will create the permissions:

Figure 9.19 – Permissions

After setting the permissions, you can see the data flow for the API:

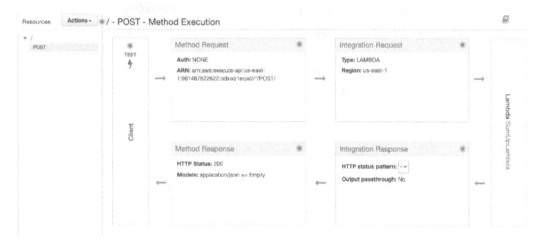

Figure 9.20 – The API flow

Now, we need to add a CORS policy. CORS is a security policy that allows a particular origin (domain or port) to browse your resource. Let's enable a CORS policy:

1. Click the **Actions** drop-down button to list the available actions, and then click **Enable CORS**:

Figure 9.21 – List of actions

2. Fill out the form and click **Enable CORS and replace existing CORS headers**. You can retain the form details as is. The form defines the following:

    A. Which methods are allowed access to the API by selecting **Methods**

    B. Which request header is required via **Access-Control-Allow-Headers**

    C. Which origins are able to call the API via **Access-Control-Allow-Origin**

    D. Gateway response types by selecting the **DEFAULT 4XX** or **DEFAULT 5XX** port. You can see the list here: `https://docs.aws.amazon.com/apigateway/latest/developerguide/supported-gateway-response-types.html`.

Figure 9.22 – Enable CORS

Congrats! You have successfully created the Lambda function and an API gateway. The next step is to test the API.

Let's test the SumUp API:

1. Click on the **Test** button in the flow:

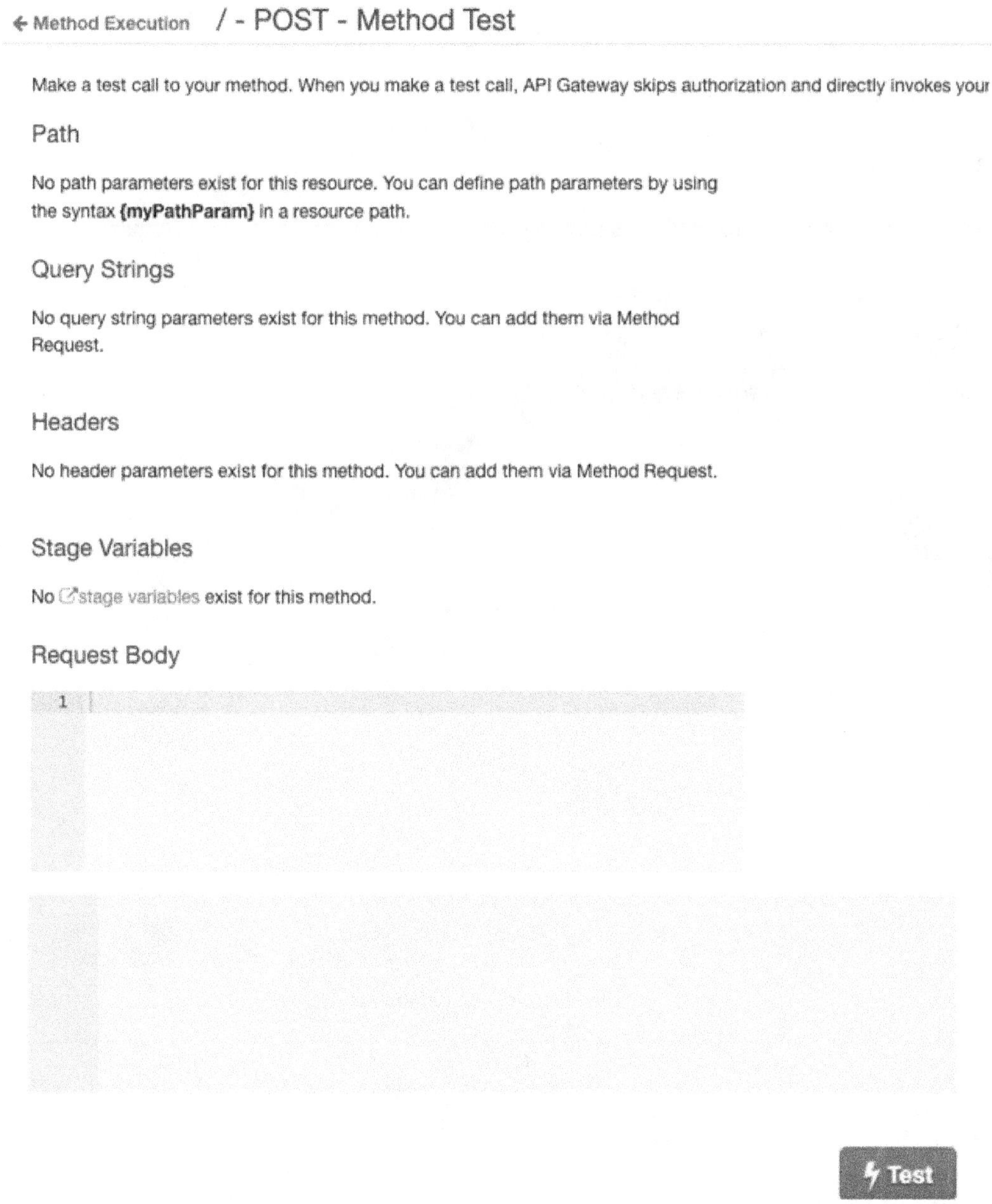

← Method Execution   / - POST - Method Test

Make a test call to your method. When you make a test call, API Gateway skips authorization and directly invokes your

Path

No path parameters exist for this resource. You can define path parameters by using the syntax **{myPathParam}** in a resource path.

Query Strings

No query string parameters exist for this method. You can add them via Method Request.

Headers

No header parameters exist for this method. You can add them via Method Request.

Stage Variables

No ☐stage variables exist for this method.

Request Body

```
1  |
```

⚡ Test

Figure 9.23 – Testing the API

2. Enter the following code in the **Request Body** field to add a parameter for Lambda:

    ```
    {
        "Number1": 10,
        "Number2": 15
    }
    ```

3. Click **Test** and see the results:

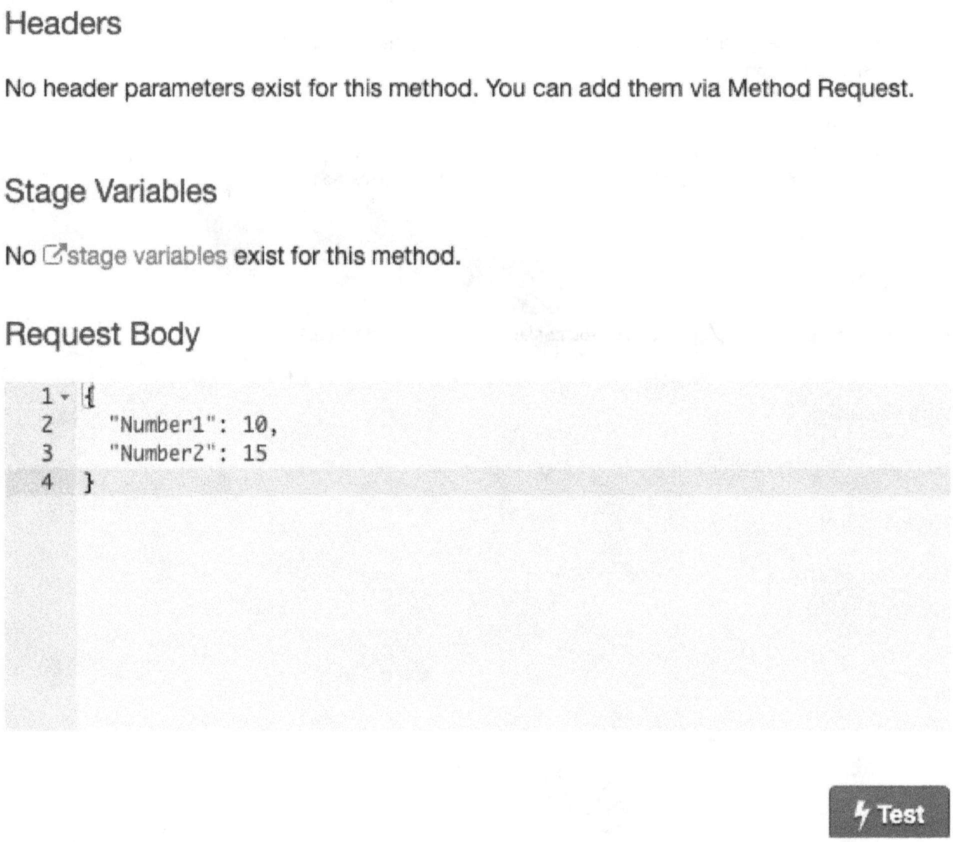

Headers

No header parameters exist for this method. You can add them via Method Request.

Stage Variables

No ☑stage variables exist for this method.

Request Body

```
1 ▾ {
2       "Number1": 10,
3       "Number2": 15
4   }
```

⚡ **Test**

Figure 9.24 – Adding a parameter

Here are the results:

```
*****************************************************************************
*******************************************************c44fd1, X-Amz-Source-Arn=arn:aws:
execute-api:us-east-1:961487522622:cdxwz1eqw2/test-invoke-stage/POST/, X-Amz-Security-Token
=IQoJb3JpZ21uX2VjEBAaCXVzLWVhc3QtMSJIMEYCIQCuLkz5BVMw/ZgjWLGFqFRO17UuYdpgwUAArCfggLMcFgIhAM
qJBxUufkOiZqBgsWkmpu8vVHUpAEwcC2sRpsiSaodLKswECBkQABoM [TRUNCATED]
Wed Nov 16 16:24:58 UTC 2022 : Endpoint request body after transformations: {
  "Number1": 10,
  "Number2": 15
}
Wed Nov 16 16:24:58 UTC 2022 : Sending request to https://lambda.us-east-1.amazonaws.com/20
15-03-31/functions/arn:aws:lambda:us-east-1:961487522622:function:SumUpLambda/invocations
Wed Nov 16 16:24:58 UTC 2022 : Received response. Status: 200, Integration latency: 397 ms
Wed Nov 16 16:24:58 UTC 2022 : Endpoint response headers: {Date=Wed, 16 Nov 2022 16:24:58 G
MT, Content-Type=application/json, Content-Length=30, Connection=keep-alive, x-amzn-Request
Id=d51191a9-1476-4c03-b4ad-12e243ec72ba, x-amzn-Remapped-Content-Length=0, X-Amz-Executed-V
ersion=$LATEST, X-Amzn-Trace-Id=root=1-63750eda-7212934a09b0f6085f4c0235;sampled=0}
Wed Nov 16 16:24:58 UTC 2022 : Endpoint response body before transformations: {"statusCod
e": 200, "Sum": 25}
Wed Nov 16 16:24:58 UTC 2022 : Method response body after transformations: {"statusCode": 2
00, "Sum": 25}
Wed Nov 16 16:24:58 UTC 2022 : Method response headers: {X-Amzn-Trace-Id=Root=1-63750eda-72
12934a09b0f6085f4c0235;Sampled=0, Content-Type=application/json}
Wed Nov 16 16:24:58 UTC 2022 : Successfully completed execution
Wed Nov 16 16:24:58 UTC 2022 : Method completed with status: 200
```

Figure 9.25 – The result of the API response

When you check the logs, you can see the results of the API response. As you can see, the sum of the values is 25.

In this topic, we implemented an API that used Python in the Lambda code. As you saw, creating an API is an easy solution in AWS. This way, you can focus on the backend implementation instead of focusing on the infrastructure.

Summary

In this chapter, we learned how to use the AWS API Gateway service and how to create an API gateway that has a backend service with Python Lambda. API Gateway is useful when you need to implement an API service with backend support via Python. It comes with scalability, logging, and monitoring advantages. In the next chapter, we will take a look at the basics of DynamoDB and NoSQL.

10

Using Python with NoSQL (DynamoDB)

In this chapter, we are going to learn how to create a NoSQL database with DynamoDB. After creating the database, we will carry out a database operation in DynamoDB using Python. **NoSQL** is a database type that is used to manage data more flexibly than a relational database. In relational databases, there are tables and predefined data types that can be used for database operations. In NoSQL, you can store JSON, raw, or key-value data, depending on the NoSQL database. Let's deep-dive into NoSQL databases.

The chapter covers the following topics:

- What is a NoSQL database?
- What is a DynamoDB database?
- DynamoDB operations with Python

What is a NoSQL database?

A NoSQL database is used to store unstructured data. The idea comes from big data; most applications and devices create data, and this data is valuable if you store and process it afterward. The volume of data is increasing day by day, and we need to store this data. Think about new cars; they have different devices to store data. We can extend our example to white goods, social media, and so on. In general, relational databases are useful for structured data and a level of records that runs into the millions. Thus, when it comes to handling millions of records as well as unstructured data, NoSQL is useful.

The following figure shows how different data sources can be generated to be stored in a NoSQL database. We have social media resources and machines in cars and planes that generate different data formats:

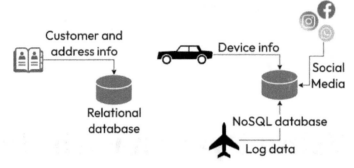

Figure 10.1 – NoSQL

There are different types of NoSQL databases.

Key-value database

In this NoSQL database type, you can access data based on keys. For example, you have customer ID as a key, and address, age, and family information as values. When you need to access the value, you just provide the key as a query parameter:

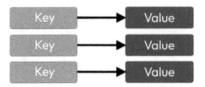

Figure 10.2 – A key-value database

A key-value database is useful and even works on billions of records. We will investigate DynamoDB, which is a key-value database, in an upcoming section.

Document database

A document database is another type of NoSQL database that can store unstructured data such as JSON. It is useful if you need to store unstructured big data and retrieve data with different parameters:

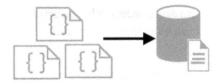

Figure 10.3 – Document database

You can see the sample JSON as follows:

```
{
    "employee": {
        "name":"Jack",
        "age":25
    }
}
```

There are other types of NoSQL databases, such as graph and column, but we won't focus on them in this book. I would recommend reading more over here: https://en.wikipedia.org/ wiki/NoSQL.

We have learned the definition of a NoSQL database and taken a look at some types of NoSQL databases. For the next step, we will focus on DynamoDB, which is one type of key-value database.

What is a DynamoDB database?

A **DynamoDB database** is a key-value NoSQL database that is managed by AWS. When you use DynamoDB, you don't need to create a new database. You don't need to provision a server either; it is fully managed by AWS. It is one of the most popular cloud-based NoSQL databases, and the performance is very good if you are using key-based access. The main advantage is that you can access data within a latency of milliseconds along with billions of records.

These are the features of DynamoDB:

- Fully managed by AWS
- Autoscaling without any configuration
- Built-in integration with other AWS services
- Supports monitoring and logging
- Supports database backup and restoration
- Pay-as-you-go model – you pay for how much you use from this service

Creating a DynamoDB database

In this subtopic, we are going to create a DynamoDB database. Let's follow the instructions step by step:

1. Type DynamoDB into the search box and click the **DynamoDB** option that appears under the **Services** section:

Figure 10.4 – Console search

2. Click **Tables** on the left side, and then click the **Create table** button:

Figure 10.5 – Create table

3. Fill out the **Table name**, **Partition key**, and **Sort key** details in order to create the table:

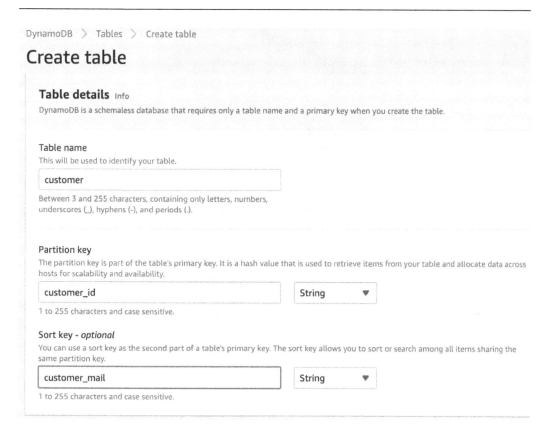

Figure 10.6 – Table details – part 1

Table name represents the name of the table. We will create a sample customer table.

Partition key is going to be used as a primary key. DynamoDB is a key-value database; hence, you can easily search for data based on the key. In this case, we will use **customer_id** as a primary key.

DynamoDB allows you to search with a sort key in addition to the partition key. We will use **customer_mail** in the **Sort key** field.

4. Scroll down and fill out **Capacity mode**, **Read capacity**, **Write capacity**, **Auto scaling**, **Local secondary indexes**, and **Global secondary indexes**. For the input, keep the following default values as is:

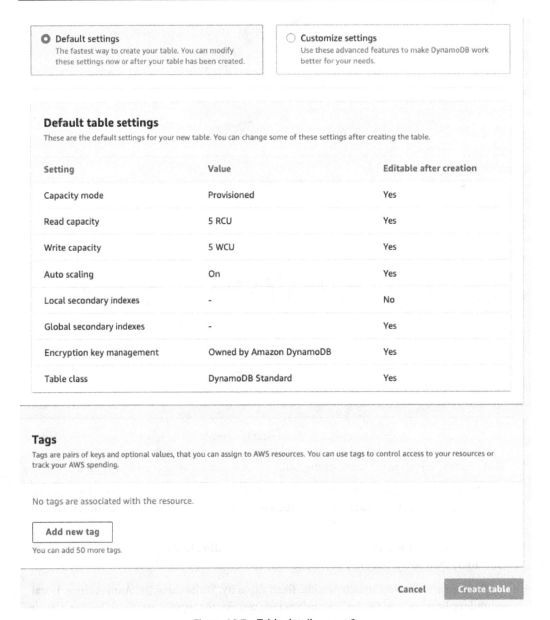

Figure 10.7 – Table details – part 2

Capacity mode defines the reserved capacity for the table. If you select the provisioned mode, AWS reserves your predefined capacity to be used by the queries. Another option is to define on-demand for unplanned capacity reservations.

Read capacity and **write capacity** define how many read and write requests are supported for this table.

Regarding **Auto scaling**, AWS manages the scaling feature.

Local secondary indexes and **Global secondary indexes** are used if you need more index values in addition to the primary key and sort key. The local secondary index allows you to create an additional index that has the same partition ID with a different sort key from the base table. You need to define this during table creation. On the other hand, a global secondary index allows you to create an index that can have a different partition ID and sort key from the base primary key.

5. Click **Create table**, as you saw in the previous screenshot, and you will see the created table in the list:

	Name ▲	Status	Partition key	Sort key	Indexes
☐	customer	✓ Active	customer_id (S)	customer_mail (S)	0

Figure 10.8 – The table list

6. Let's insert one of the items via the AWS Management Console. Select **customer** under the **Tables** list:

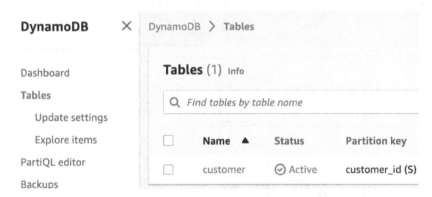

Figure 10.9 – Customer table

You will see the details of the table:

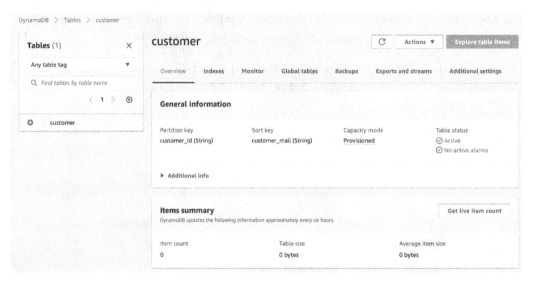

Figure 10.10 – Table details

7.   Click the **Actions** drop-down button and select **Create item**:

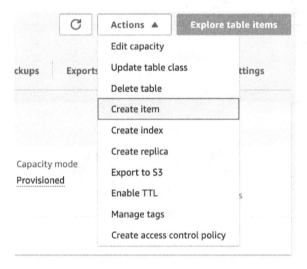

Figure 10.11 – Create item

8.   After clicking this, you will see an item creation page, titled **Create item**. You can fill out a form or insert the JSON directly. In this example, we will insert the code via **JSON view**. DynamoDB creates a template for you:

Figure 10.12 – The JSON view

Paste the following JSON as an example:

```
{
    "customer_id": {
        "S": "123"
    },
    "customer_mail": {
        "S": "serkansakinmaz@gmail.com"
    },
    "name": {
        "S": "Serkan"
    },
    "address": {
        "S": "Germany"
    }
}
```

The JSON is simple and consists of customer_id, customer_mail, name, and address information.

9.   Click **Create item**:

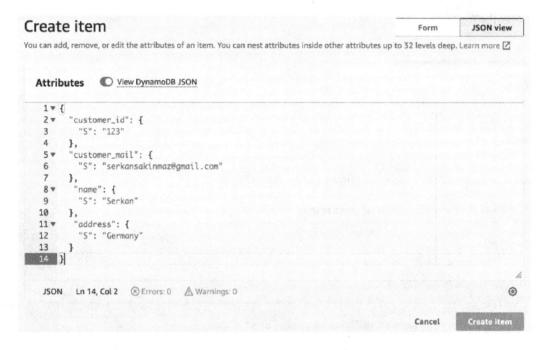

Figure 10.13 – Creating an item

After the creation, you will be forwarded to the **Tables** page:

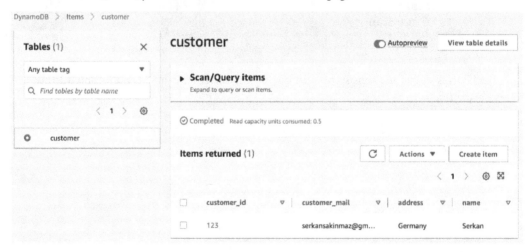

Figure 10.14 – The item list

Since you are using NoSQL, you can also insert the JSON, which is a different format from the previous JSON that we inserted. The following JSON is also valid for the customer table:

```
{
    "customer_id": {
      "S": "1234"
    },
    "customer_mail": {
      "S": "jane@gmail.com"
    },
    "name": {
      "S": "Jane"
    },
    "profession": {
      "S": "Data Engineer"
    }
}
```

As you see, we have removed the `address` field and added `profession` as a new field without any issue.

In this section, we have created a DynamoDB table and inserted data via the console. As you can see, DynamoDB is a key-value database and you can insert different JSON formats, which provides flexibility.

DynamoDB operations with Python

In this section, we are going to read the DynamoDB table via Python. To execute a Python function, we will implement a Lambda function to read data from DynamoDB. Carry out the following steps:

1. We will create the required permissions to allow Lambda to read from DynamoDB. Open IAM and click **Policies** on the left-hand side:

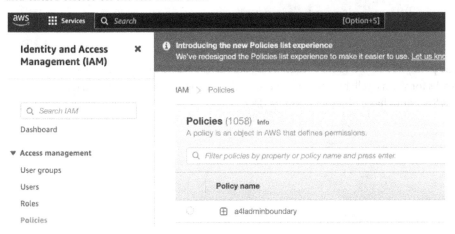

Figure 10.15 – IAM policies

2. Click **Create policy**:

Figure 10.16 – Creating a policy

3. Paste the following policy:

```
{
    "Version": "2012-10-17",
    "Statement": [
        {
            "Effect": "Allow",
            "Action": [
                "dynamodb:BatchGetItem",
                "dynamodb:GetItem",
                "dynamodb:Query",
                "dynamodb:Scan",
                "dynamodb:BatchWriteItem",
                "dynamodb:PutItem",
                "dynamodb:UpdateItem"
            ],
            "Resource": "arn:aws:dynamodb:us-east-1:961487522622:table/customer"
        }
    ]
}
```

The policy allows you to read from the DynamoDB table. In general, the following access policy works for you as well; however, you need to change the account ID that you have, because every AWS account has a different account ID:

Create policy

A policy defines the AWS permissions that you can assign to a user, group, or role. You can create and edit a policy in the visual editor and using

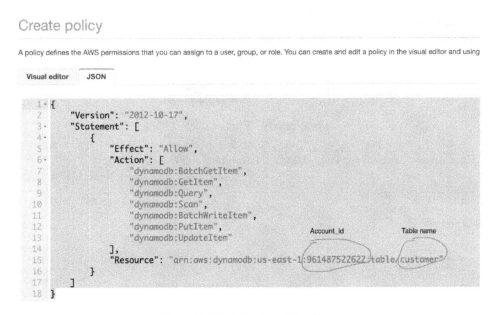

Figure 10.17 – A DynamoDB policy

4. You can add the policy name and finish creating the policy. In this example, I am using **DynamoDBCustomerTableOperations** as a policy name:

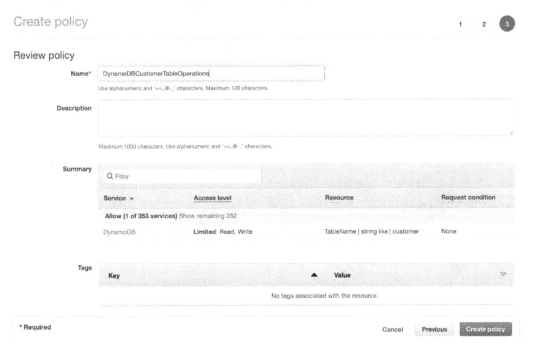

Figure 10.18 – Policy creation

5. We now need to create a role. This role will be attached to Lambda to access DynamoDB. Click **Create role** in the IAM service:

Figure 10.19 – The IAM role

6. Since we need a policy for Lambda, select **Lambda** in the **Use case** section:

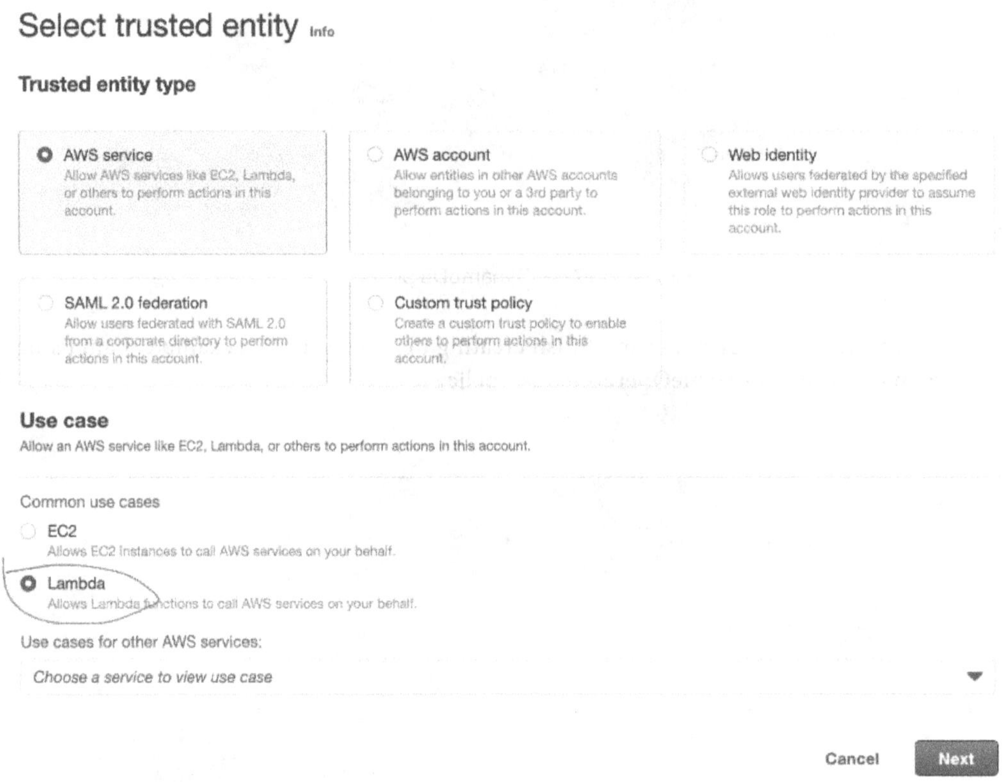

Select trusted entity Info

Trusted entity type

○ **AWS service**
Allow AWS services like EC2, Lambda, or others to perform actions in this account.

○ **AWS account**
Allow entities in other AWS accounts belonging to you or a 3rd party to perform actions in this account.

○ **Web identity**
Allows users federated by the specified external web identity provider to assume this role to perform actions in this account.

○ **SAML 2.0 federation**
Allow users federated with SAML 2.0 from a corporate directory to perform actions in this account.

○ **Custom trust policy**
Create a custom trust policy to enable others to perform actions in this account.

Use case
Allow an AWS service like EC2, Lambda, or others to perform actions in this account.

Common use cases

○ EC2
Allows EC2 instances to call AWS services on your behalf.

● Lambda
Allows Lambda functions to call AWS services on your behalf.

Use cases for other AWS services:

Choose a service to view use case ▼

Cancel   Next

Figure 10.20 – The IAM role for Lambda

7. As depicted in the following screenshot, add the policy that we created to access Lambda:

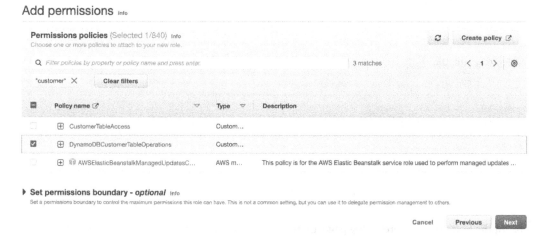

Figure 10.21 – Selecting the policy

8. Fill in **Role name** and create the role. As you see, the name we have given to the Lambda function is **DynamoDBCustomerTableRole**. Scroll down and click the **Create role** button:

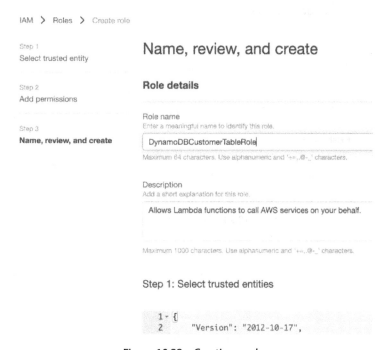

Figure 10.22 – Creating a role

9. The **Create function** page opens up. We create a Lambda function by adding `readFromDynamoDB` to **Function name** and **Python 3.9** to **Runtime**:

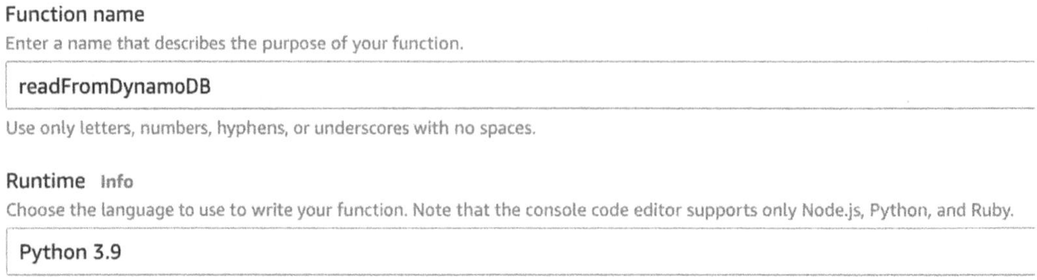

Function name

Enter a name that describes the purpose of your function.

readFromDynamoDB

Use only letters, numbers, hyphens, or underscores with no spaces.

Runtime Info

Choose the language to use to write your function. Note that the console code editor supports only Node.js, Python, and Ruby.

Python 3.9

Figure 10.23 – Creating a function

10. At the bottom of the preceding page, there is a panel to define the execution policy. Select **Use an existing role** under the **Execution role** section and select the role that we created:

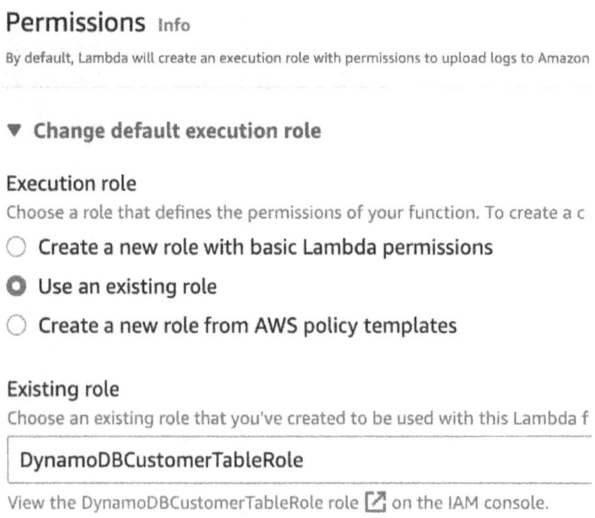

Permissions Info

By default, Lambda will create an execution role with permissions to upload logs to Amazon

▼ Change default execution role

Execution role

Choose a role that defines the permissions of your function. To create a c

○ Create a new role with basic Lambda permissions

● Use an existing role

○ Create a new role from AWS policy templates

Existing role

Choose an existing role that you've created to be used with this Lambda f

DynamoDBCustomerTableRole

View the DynamoDBCustomerTableRole role 🗗 on the IAM console.

Figure 10.24 – Selecting the role

11. Lambda is ready to fill out a code block:

Figure 10.25 – The Lambda function

Paste the following code into the Lambda function:

```
import json
import boto3

def lambda_handler(event, context):

    dynamodb = boto3.resource('dynamodb', region_name="us-east-1")
    table = dynamodb.Table('customer')
    response = table.get_item(Key={'customer_id': "123", 'customer_
mail': "serkansakinmaz@gmail.com"})
    item = response['Item']
    print(item)



    return {
        'statusCode': 200,
        'body': json.dumps('Hello from Lambda!')
    }
```

The code imports the boto3 library, which provides useful functions for DynamoDB operations. boto3 is a library that includes AWS service-specific features to facilitate the implementation of cloud applications while working with Python on AWS. You can get more details from the following link: https://boto3.amazonaws.com/v1/documentation/api/latest/index.html.

As a first step, we define the `dynamodb` resource by calling the `boto3.resource` function. After calling that, we define the table that we read; it is the `dynamodb.Table` table. Once you define the table, the `table.get_item` function takes the primary key and sort key as a parameter and returns the query results.

Once you run the Lambda function, you are able to see the result:

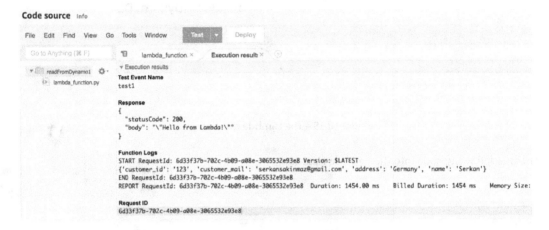

Figure 10.26 – Execution results

Congratulations! You are able to define the role and retrieve an item from Lambda. As you can see, AWS requires some configuration to access data in DynamoDB.

Summary

In this chapter, we learned about the AWS DynamoDB service and how to create a DynamoDB database in AWS. After creating the database, we implemented a Lambda Python code snippet that read items from DynamoDB. You now also know how to extend the Lambda code to insert data into a DynamoDB table. DynamoDB is useful when you need to implement a key-value database that is managed by AWS. It comes with scalability, logging, and monitoring advantages. In the following chapter, we will take a look at the Glue service.

11

Using Python with Glue

In this chapter, we are going to learn how to create a data integration pipeline with AWS Glue. **AWS Glue** is a data integration service that is used for the **Extract, Transform, and Load** (**ETL**) process. Glue is a serverless data integration service; therefore, you don't need to create and manage a server, as the infrastructure is managed by AWS. With Glue, you can collect data from different data sources, such as S3, databases, or filesystems, to process and transform the data. The result is stored in S3 or the database, or you can call an API.

The chapter covers the following topics:

- What is the AWS Glue service?
- AWS Glue service creation
- Creating a simple Python application with AWS Glue

What is the AWS Glue service?

AWS has more than 100 services. When you integrate data between AWS and other sources, you might need to load data from the source, manipulate it with some transformations, and store it in a service. AWS Glue meets these requirements and provides a service that allows the preparation of data. In the following figure, you can see a very high-level overview of Glue. As you can see, Glue extracts the data from different sources, carries out some transformation, and loads the data in another source:

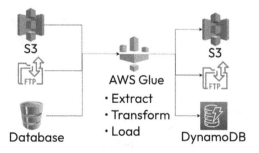

Figure 11.1 – AWS Glue

For example, let us assume you have data in S3 that is loaded by a batch process. To make it searchable, you have a requirement to store it in DynamoDB. Between these processes, one requirement is to filter, clean, and manipulate the data with some transformations. For that requirement, AWS Glue is a good option for data integration with some data manipulation.

Features of AWS Glue

AWS Glue has the following features:

- It automatically scales based on the transformation workload.

- It has wider integration with other services to load data, such as S3, RDS, and DynamoDB. Hence, you can easily read data with these services.

- You can schedule the pipeline; hence, the ETL process can be performed regularly based on the scheduled time.

- It has a data catalog feature that allows you to store metadata information for the data structure.

- It is able to generate code for the ETL pipeline. For example, you need to read CSV data from S3 to load another S3 location in JSON format. Glue automatically generates the code.

- There is Git integration, so you can easily pull code from Git to run the ETL pipeline.

- It provides a visual interface with a drag-and-drop code implementation feature.

In this section, we looked at AWS Glue's features. To understand them better, we are going to convert a CSV file to JSON using the AWS Glue service.

Creating an S3 sample file

In this section, we are going to create a simple S3 bucket that stores a CSV file. Let's follow the instructions step by step:

1. Open the AWS S3 service.

2. Click the **Create bucket** button:

Figure 11.2 – Create bucket

3. Give a unique **bucket name** and click **Create bucket** at the end of the panel:

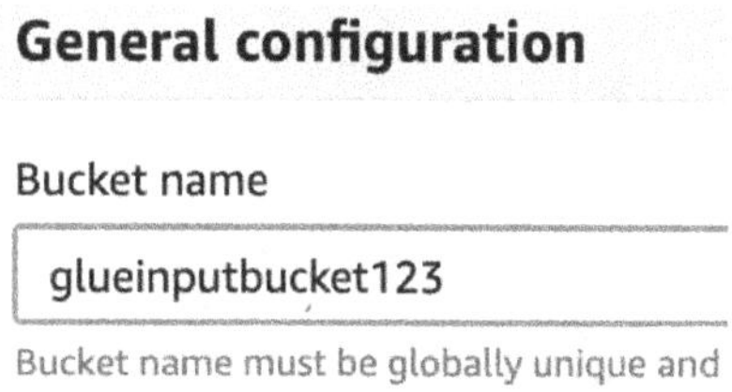

General configuration

Bucket name

glueinputbucket123

Bucket name must be globally unique and

Figure 11.3 – Input bucket

The bucket is created:

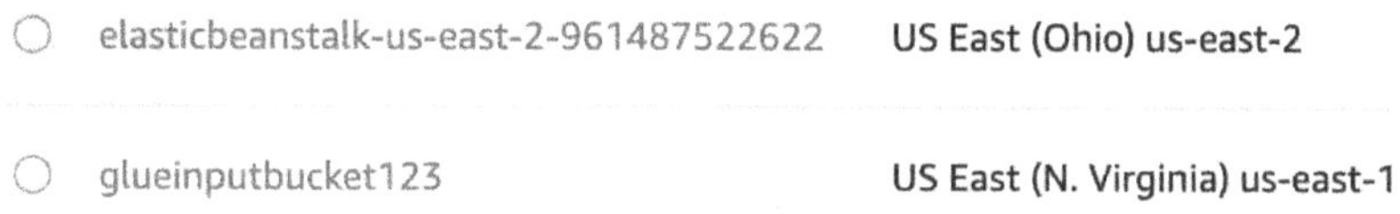

○ elasticbeanstalk-us-east-2-961487522622   **US East (Ohio) us-east-2**

○ glueinputbucket123   **US East (N. Virginia) us-east-1**

Figure 11.4 – Bucket list

4. Create an `addresses.csv` file on your computer with the following content and upload it
    to the S3 bucket. Please save the file in the UTF-8 format; otherwise, there might be an issue
    in some Glue versions:

```
id,location_id,address_1,city,state_province
1,1,2600 Middlefield Road,Redwood City,CA
2,2,24 Second Avenue,San Mateo,CA
3,3,24 Second Avenue,San Mateo,CA
4,4,24 Second Avenue,San Mateo,CA
5,5,24 Second Avenue,San Mateo,CA
6,6,800 Middle Avenue,Menlo Park,CA
7,7,500 Arbor Road,Menlo Park,CA
8,8,800 Middle Avenue,Menlo Park,CA
9,9,2510 Middlefield Road,Redwood City,CA
10,10,1044 Middlefield Road,Redwood City,CA
```

5. Click the **Upload** button within the bucket to upload the content:

Figure 11.5 – Uploading the CSV

After the upload, the bucket will include the CSV file:

Figure 11.6 – S3 content

We have successfully uploaded the file. In the next step, we will create the required permissions in order to create a Glue job.

Defining the permissions for a Glue job

In this section, we are going to define the required permissions for a Glue job:

1. Open the AWS IAM service.
2. Click **Roles** on the left-hand side:

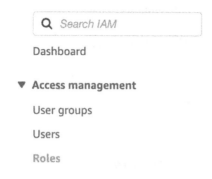

Figure 11.7 – List of IAM services

3. Click **Create role**:

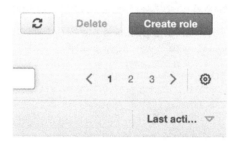

Figure 11.8 – Create role

4. Under **Use case**, select **Glue**:

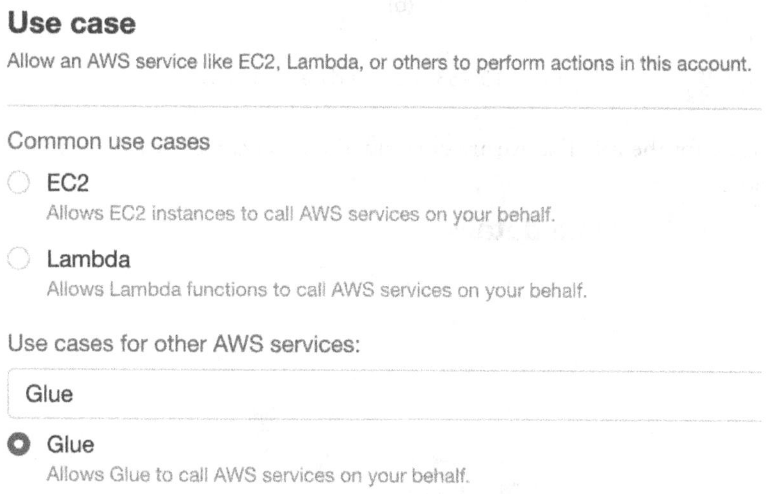

Use case

Allow an AWS service like EC2, Lambda, or others to perform actions in this account.

Common use cases

○ EC2
    Allows EC2 instances to call AWS services on your behalf.

○ Lambda
    Allows Lambda functions to call AWS services on your behalf.

Use cases for other AWS services:

    Glue

● Glue
    Allows Glue to call AWS services on your behalf.

Figure 11.9 – Selecting Glue

5. On the next page, select **AmazonS3FullAccess** and **CloudWatchFullAccess** under **Policy name**:

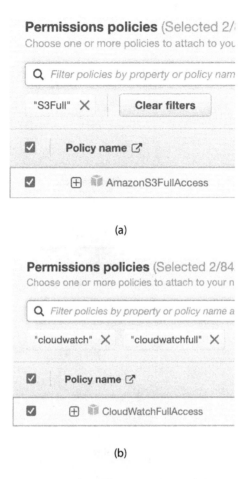

(a)

(b)

Figure 11.10 – S3 and CloudWatch access

6. Give a name for the role that we are creating, then you can click **Create role** to finish the role creation:

(a)

(b)

Figure 11.11 – Role name and creation

We have created the required role for an AWS Glue job. In the next step, we will create a simple AWS Glue job by using roles and the S3 bucket.

Creating an AWS Glue service

In this section, we are going to create an AWS Glue service. Let's follow the instructions step by step:

1. Type AWS Glue in the AWS Management Console search bar and click the **AWS Glue** result that appears:

Figure 11.12 – Console search

2. Click **Jobs** on the left-hand side:

Figure 11.13 – Glue job

3. In the **Create job** section, select **Visual with a source and target**. This will create a visual interface and predefined script in order to convert from a CSV to a JSON file:

Figure 11.14 – Create job

4. After clicking **Create** on the right side of the panel, you will see the visual editor:

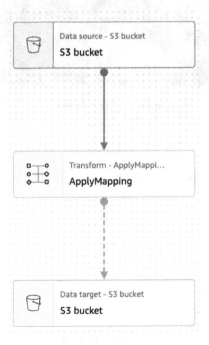

Figure 11.15 – Visual editor

5. Click **S3 bucket** under **Data source - S3 bucket** and you will see the data source details on the right side. It consists of some details on how to parse the source data. As you can see in the following figure, we set **S3 location** as a data path, **Data format** is **CSV**, and **Delimiter** is comma-separated:

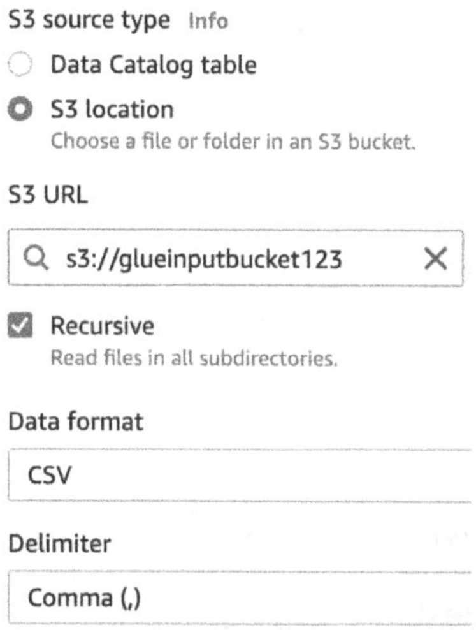

S3 source type   Info

○ Data Catalog table

● S3 location
Choose a file or folder in an S3 bucket.

S3 URL

🔍 s3://glueinputbucket123   ✕

☑ Recursive
Read files in all subdirectories.

Data format

CSV

Delimiter

Comma (,)

Figure 11.16 – Data source

6. Select the **Transform** tab from the panel and you will see the following data mapping. This mapping is generated by Glue:

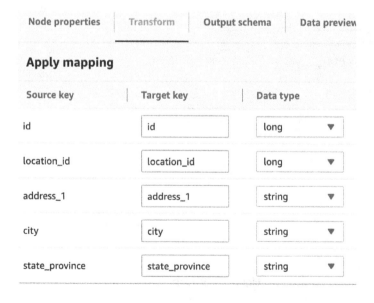

Figure 11.17 – Mapping

7. Select the **Data target properties - S3** tab from the panel and fill out the panel with target details. Since we are converting to JSON, the format will be **JSON**. The target location could also be another S3 bucket; in this example, I will give the same S3 location for input and output:

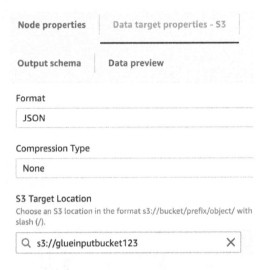

Figure 11.18 – Data target

8. Select the **Job details** tab in order to fill out other information such as the job name and script. You can see these details in *steps 9* and *10*:

Figure 11.19 – Job details

9. Fill in the job's **Name** and **IAM Role** fields to run the Glue job:

Basic properties Info

Name

job1

Description - *optional*

IAM Role

Role assumed by the job with permission to access your da
targets, temporary directory, scripts, and any libraries used

roleforglue

Figure 11.20 – Name and role

10. There is one more configuration left. Scroll down and fill in the **Script filename** and **Script path** details that Glue will create:

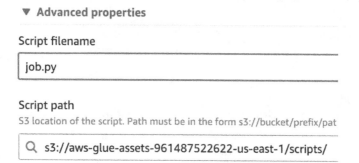

▼ **Advanced properties**

Script filename

job.py

Script path

S3 location of the script. Path must be in the form s3://bucket/prefix/pat

🔍 s3://aws-glue-assets-961487522622-us-east-1/scripts/

Figure 11.21 – Script filename and path

11. Click **Save**. As you can see, Glue has created a Python Spark script that is going to convert CSV to JSON. **PySpark** is a data processing library that can also be used in the AWS Glue job:

Figure 11.22 – Code generation

12. Click **Run** on the right side of the panel:

Figure 11.23 – Button panel for Run

After some time, you can check the job status from the **Runs** tab:

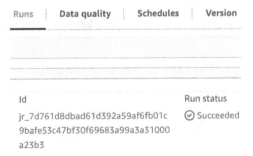

Figure 11.24 – Runs tab

When you check the S3 folder, the file is converted to JSON. Here is some sample output:

```
{"id":"1","location_id":"1","address_1":"2600 Middlefield
Road","city":"Redwood City","state_province":"CA"}
{"id":"2","location_id":"2","address_1":"24 Second
Avenue","city":"San Mateo","state_province":"CA"}
{"id":"3","location_id":"3","address_1":"24 Second
Avenue","city":"San Mateo","state_province":"CA"}
```

Congrats! You are able to convert a CSV file to a JSON file. As you can see, AWS Glue created a predefined script to make some ETL jobs.

Summary

In this chapter, we learned about the AWS Glue service and how to create an ETL pipeline with AWS Glue. Glue is very efficient when you need to create data pipelines. One cool feature of Glue is the visual flow generator, which allows you to create a flow with drag and drop. It makes it easy to create and generate the flow, which saves lots of time. In addition to that, for people who don't have that much code experience, Glue's visual flow facilitates their tasks. Hence, if you work with data, Glue is one of the best services within AWS. In the next chapter, we will create a sample project within AWS using the Python programming language.

12

Reference Project on AWS

In this chapter, we are going to create a sample application with Python on AWS. This is the final chapter of the book. We have learned about different AWS services and implemented sample Python applications with these services. In this chapter, we will use multiple services to create an end-to-end Python application.

The chapter covers the following topics:

- What have we learned?
- Introducing the end-to-end Python application
- The coding of the Python application

What have we learned?

AWS has more than a hundred services, and we have learned about the important Python-related services. Let's walk through those services:

- **Lambda**: Lambda is a cloud computing service that allows you to run Python applications. You don't need to provision any server; Lambda manages the infrastructure.

- **EC2**: EC2 provides a server machine in the cloud. You can create a server and install the required applications, or whatever you want.

- **Elastic Beanstalk**: Elastic Beanstalk is used to deploy Python-based web applications.

- **CloudWatch**: CloudWatch is a logging and monitoring service on AWS. You can easily track your services.

- **RDS**: RDS is a relational database service on AWS. If you need a database, you can easily create it without managing the server.

- **API Gateway**: API Gateway is used to create, maintain, and publish an application programming interface.

- **DynamoDB**: DynamoDB is a key-value database that is used to query and store billions of records on AWS. It is also a **NoSQL database**.

- **AWS Glue**: AWS Glue is a data integration service that is used for ETL.

Introducing the Python application

Let us understand the high-level architecture of the Python application:

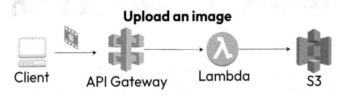

Figure 12.1 – Project architecture

The application collects images to be stored in S3 buckets. The API gateway is used for integration between clients and the Lambda service. Lambda retrieves the information and puts data into S3.

The coding of the Python application

Let's implement the application step by step.

Creating S3 buckets to store images

In this subsection, we are going to create an S3 bucket to hold images, which is uploaded via API Gateway. S3 will store the image and provide it whenever requested:

1. Create a bucket and click the **Create bucket** button at the bottom of the page:

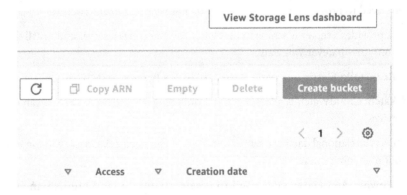

Figure 12.2 – An S3 bucket

2. We filled in the **Bucket name** field as `python-book-image`; you can use whatever you want. After adding the **bucket name**, click **Create bucket** to create a new bucket:

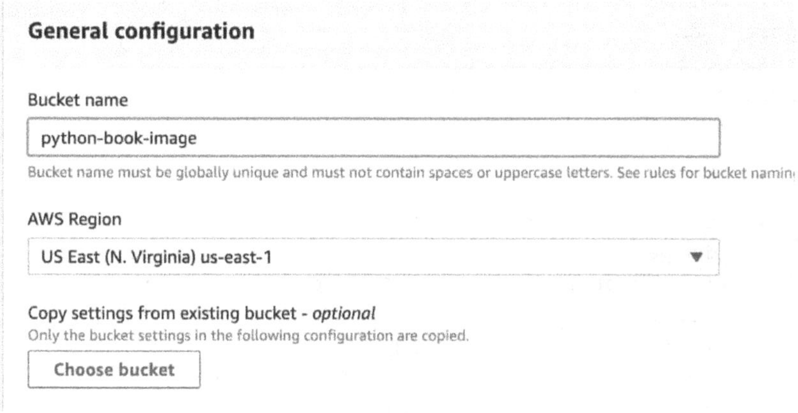

General configuration

Bucket name

python-book-image

Bucket name must be globally unique and must not contain spaces or uppercase letters. See rules for bucket naming

AWS Region

US East (N. Virginia) us-east-1   ▼

Copy settings from existing bucket - *optional*
Only the bucket settings in the following configuration are copied.

Choose bucket

Figure 12.3 – Bucket configuration

We have created an S3 bucket.

Creating Lambda code

In this subsection, we are going to implement a Lambda code that accepts the image upload request from API Gateway and stores the image in the S3 bucket:

1. Create a Lambda function via the AWS Management Console. You can see the **Function name** field of the Lambda function and **Runtime** in the following screenshot within the Lambda creation step:

Basic information

Function name
Enter a name that describes the purpose of your funct

UploadImageToS3

Use only letters, numbers, hyphens, or underscores wi

Runtime Info
Choose the language to use to write your function. No

Python 3.9

Figure 12.4 – The Lambda function

2. Paste the following code to the Lambda code source:

```
import boto3
import base64
import json

def lambda_handler(event, context):
    try:
        s3 = boto3.resource('s3')
        s1 = json.dumps(event)
        data = json.loads(s1)
        image = data['image_base64']
        file_content = base64.b64decode(image)
        bucket = data['bucket']
        s3_file_name = data['s3_file_name']
        obj = s3.Object(bucket,s3_file_name)
        obj.put(Body=file_content)
        return 'Image is uploaded to ' + bucket
    except BaseException as exc:
        return exc
```

3. Once pasted, deploy the Lambda function by clicking the **Deploy** button:

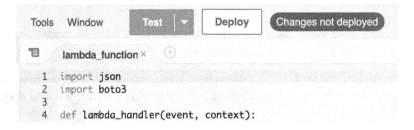

Figure 12.5 – Lambda deployment

Let's take a look at the code details. First, we import the json, base64, and boto3 libraries. The json library is used to parse data, which comes in JSON format, and boto3 is used to upload files to S3 as well as generate a URL for retrieving the file. In addition to that, base64 is used to decode and encode the image.

The following lines of code are parsing the parameters and decoding the contents of the image to store S3. Hence, we can use the bucket name and S3 filename. The bucket name is represented as bucket in the code and the S3 filename is represented as s3_file_name:

```
s1 = json.dumps(event)
data = json.loads(s1)
image = data['image_base64']
```

```
file_content = base64.b64decode(image)
bucket = data['bucket']
s3_file_name = data['s3_file_name']
```

Once we have parameters, we can use the `boto3` library to upload the file from local to S3:

```
obj = s3.Object(bucket,s3_file_name)
obj.put(Body=file_content)
```

We have implemented the code for the application. In order to run this code, we have to create permissions, the steps for which are explained in the next subsection.

Creating permissions for the services

We are now going to create permissions to upload a file to S3 and call a Lambda function from API Gateway:

1. Open the IAM role and create a new role for **Lambda**:

Use case

Allow an AWS service like EC2, Lambda, or others to perform actions in this account.

Common use cases

○ EC2
Allows EC2 instances to call AWS services on your behalf.

● Lambda
Allows Lambda functions to call AWS services on your behalf.

Figure 12.6 – Creating a role

2. Select **AmazonS3FullAccess** and **CloudWatchFullAccess** from the list:

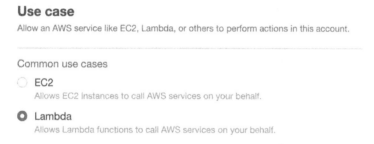

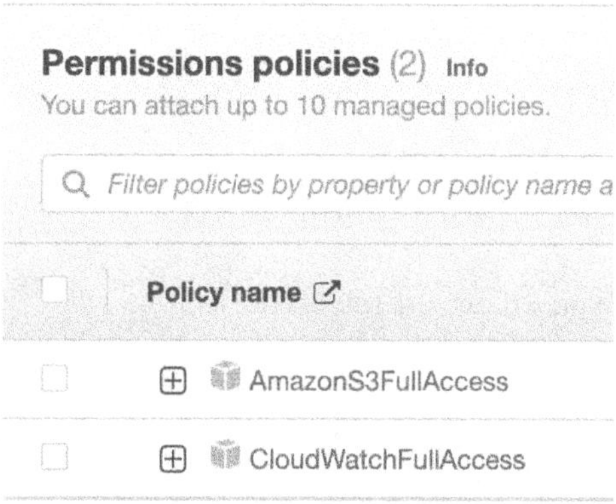

Figure 12.7 – Adding policies

3. Click the **Next** button:

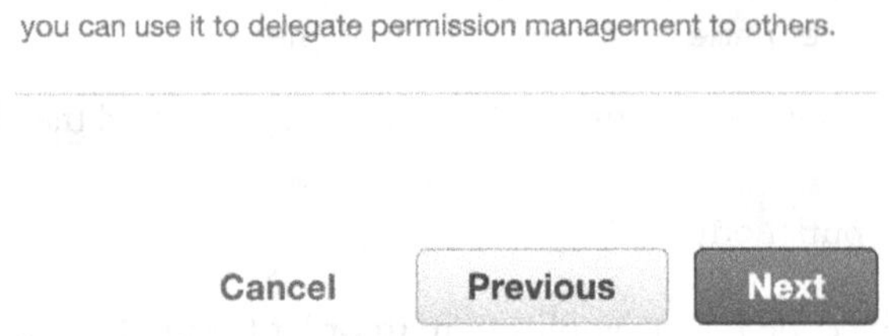

you can use it to delegate permission management to others.
Cancel
Previous
Next

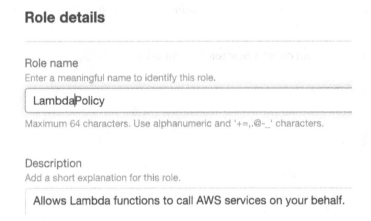

Figure 12.8 – Adding policies

4. Add the role name:

Role details
Role name
Enter a meaningful name to identify this role.
LambdaPolicy
Maximum 64 characters. Use alphanumeric and '+=,.@-_' characters.
Description
Add a short explanation for this role.
Allows Lambda functions to call AWS services on your behalf.

Figure 12.9 – Naming the role

5. Complete creating the role by clicking the **Create role** button:

Cancel
Previous
Create role

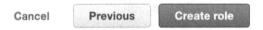

Figure 12.10 – Create role

6. After creating the role, you will see the role on the list:

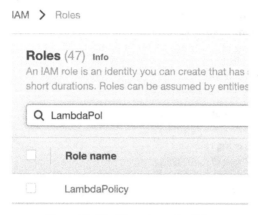

Figure 12.11 – The role on the list

In this subsection, we have created a role to be used in the Lambda function to execute the code. Let's attach the role to the Lambda function.

Attaching the role to the Lambda function

We are now going to add permissions to the Lambda function:

1. Open the Lambda function and click **Permissions** under the **Configuration** tab:

Figure 12.12 – Lambda permissions

2. Edit the permissions and select **LambdaPolicy** from the existing role. This role was created in the previous subsection:

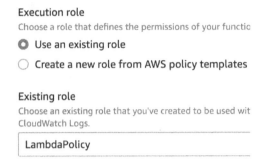

Figure 12.13 – Attaching the permission

With this configuration, Lambda is able to execute the code. It is time to start implementing API Gateway, which will use a Lambda function as a backed function.

Creating an API gateway to upload the image

In this step, we are going to create an API gateway to upload the image:

1. Open the API Gateway service and create a REST API:

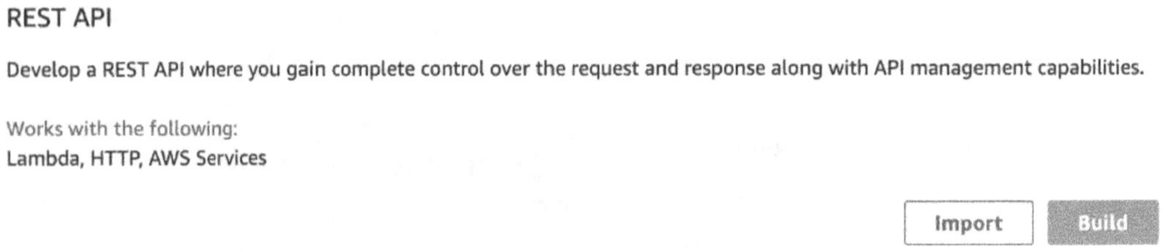

REST API

Develop a REST API where you gain complete control over the request and response along with API management capabilities.

Works with the following:
Lambda, HTTP, AWS Services

Figure 12.14 – Creating a REST API

2. Provide a name for the REST API. We will use the name `UploadImageToS3` in this subsection:

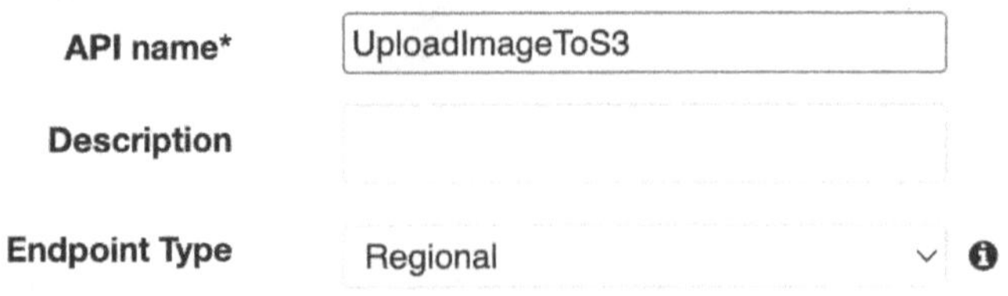

API name*	UploadImageToS3
Description	
Endpoint Type	Regional

Figure 12.15 – Naming the REST API

3. In the **Actions** drop-down list, click **Create Method**:

Figure 12.16 – Creating a method

4. Select **POST** from the available options:

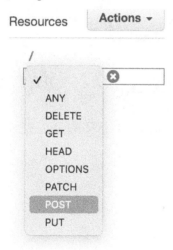

Figure 12.17 – The POST method

5. We will use **Lambda Function** as the integration type and scroll down to click **Save**:

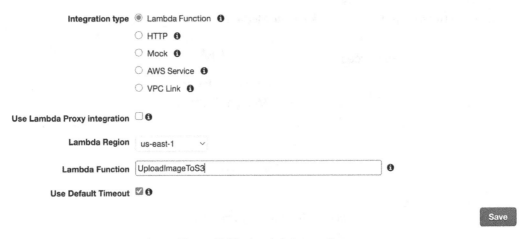

Figure 12.18 – Lambda integration

6.   The API is ready to use. Enable the CORS policy as we explained in *Chapter 9*, then click **Deploy API** in the **Actions** drop-down list:

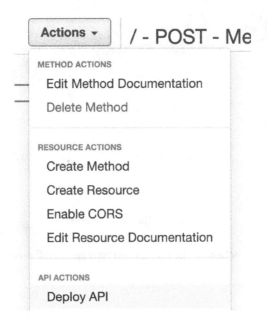

Figure 12.19 – Deploying the API

7.   We are ready to deploy the API. Add a stage name and click **Deploy**:

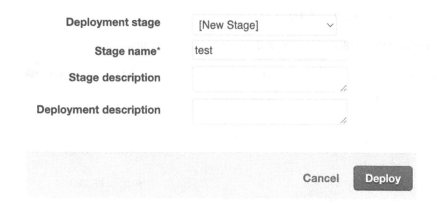

Figure 12.20 – Naming the stage

8. In the **Export** tab, there are multiple alternatives to call the API. We will use Postman to call the API. **Postman** is a platform that allows you to build and test the API. For this application, you can also test another platform such as **Swagger**. Postman is an easy way to use and test an API. In the following subsection, we will explain how to download and use it. Since it is simpler in terms of installation and use, I will proceed with Postman.

    Select the **Export as Swagger + Postman Extensions** icon; you can export and download either the JSON or YAML format:

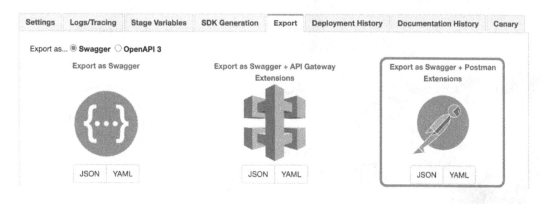

Figure 12.21 – Exporting the API

This file will be used in Postman to test the API.

Using Postman to test the API

We have completed the implementation. In this step, we are going to test the API via Postman:

1. Download and install Postman from the following website: `https://www.postman.com/`.

2. In the Postman application, click the **Import** button:

Figure 12.22 – Importing the API

3. Select the JSON file that we downloaded within API Gateway and click **Open**:

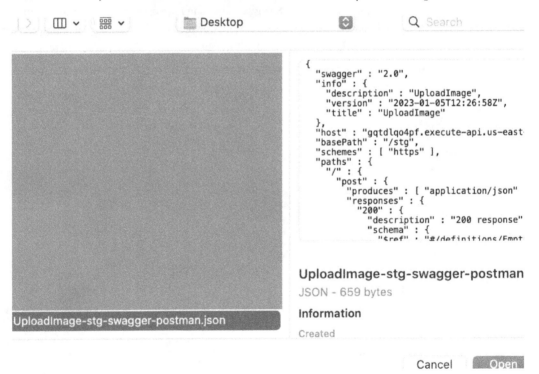

Figure 12.23 – Importing the JSON

4. You will see confirmation of the API. Click **Import** as a final step:

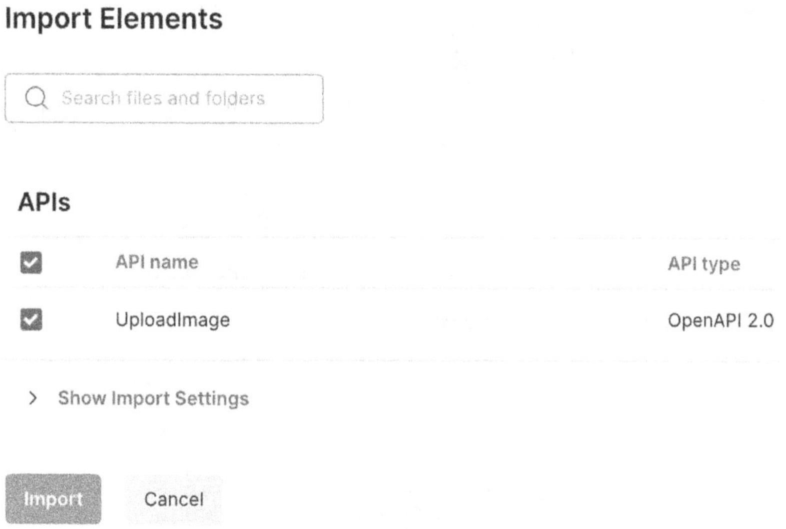

Import Elements

Q Search files and folders

APIs

	API name	API type
☑	API name	API type
☑	UploadImage	OpenAPI 2.0

> Show Import Settings

Import   Cancel

Figure 12.24 – Import the JSON

5. Once you have imported the API, you are ready to call the API. In the **POST** section, select the **raw** request type with **JSON** as follows:

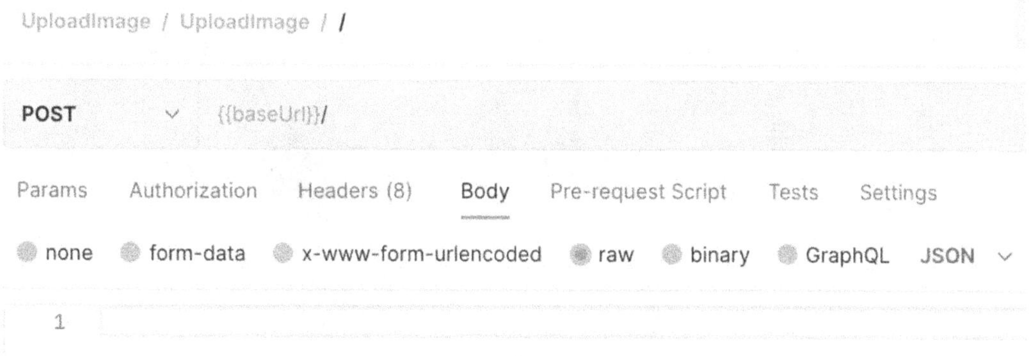

UploadImage / UploadImage / **/**

POST   ∨   {{baseUrl}}/

Params   Authorization   Headers (8)   **Body**   Pre-request Script   Tests   Settings

● none   ● form-data   ● x-www-form-urlencoded   ● raw   ● binary   ● GraphQL   JSON ∨

1

Figure 12.25 – The raw parameter

6. Paste the following JSON to call the API:

```
{
    "image_base64":"iVBORw0KGgoAAAANSUhEUgAAAAEAAAABCAQAAAC1H
    AwCAAAAC01EQVR42mNk+A8AAQUBAScY42YAAAAASUVORK5CYII=",
    "bucket":"python-book-image",
```

```
            "s3_file_name":"image.jpeg"
    }
```

Let's break down the JSON file:

- `image_base64` represents the `base64` code of a sample image that is going to be saved to the S3 bucket. You can also convert a sample image to `base64` code with libraries and online converters.

- The `bucket` parameter represents the location of the S3 bucket.

- `s3_file_name` represents the name and extension of the content.

This can be seen in the following screenshot:

Figure 12.26 – Request JSON

7. Click the **Send** button in order to call the API. Once you click it, you can see the response of the API:

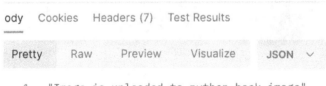

Figure 12.27 – JSON response

We have successfully called the API. Let's check with the S3 bucket whether the image is uploaded.

8. Open the `python-book-image` S3 bucket and see the uploaded jpeg file:

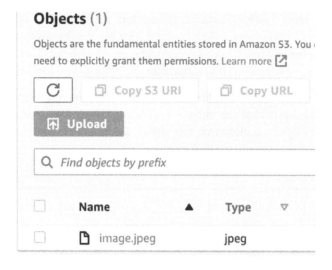

Figure 12.28 – S3 content

9. Download the file and check the sample image. When you download it, you will see a very small point. You can make it bigger by clicking the + magnifying glass icon on your image viewer to see it clearly:

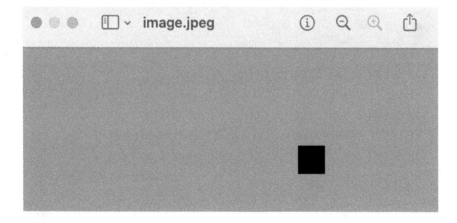

Figure 12.29 – The image

Congratulations! You have successfully uploaded the image using API Gateway, Lambda, and S3 services.

Summary

In this chapter, we have created an application to upload an image using API Gateway, Lambda, and S3. The image is converted to `base64` to be stored in S3. One of the best aspects of using Lambda, S3, and API Gateway is that we haven't provisioned any server. Lambda, S3, and API Gateway are serverless and we don't need to manage the infrastructure. AWS manages and handles it for you.

We have finished all the chapters and learned how to use the most common AWS services with Python. I hope all the chapters have provided you with good knowledge about AWS. Following this, you can implement more complex Python projects with these services as well as use more services within AWS.

Index

Packtpub.com

Subscribe to our online digital library for full access to over 7,000 books and videos, as well as industry leading tools to help you plan your personal development and advance your career. For more information, please visit our website.

Why subscribe?

- Spend less time learning and more time coding with practical eBooks and Videos from over 4,000 industry professionals

- Improve your learning with Skill Plans built especially for you

- Get a free eBook or video every month

- Fully searchable for easy access to vital information

- Copy and paste, print, and bookmark content

Did you know that Packt offers eBook versions of every book published, with PDF and ePub files available? You can upgrade to the eBook version at packtpub.com and as a print book customer, you are entitled to a discount on the eBook copy. Get in touch with us at customercare@packtpub.com for more details.

At www.packtpub.com, you can also read a collection of free technical articles, sign up for a range of free newsletters, and receive exclusive discounts and offers on Packt books and eBooks.

Other Books You May Enjoy

If you enjoyed this book, you may be interested in these other books by Packt:

Industrializing Financial Services with DevOps

https://packt.link/9781804614341

Spyridon Maniotis

ISBN: 978-1-80461-434-1

- Understand how a firm's corporate strategy can be translated to a DevOps enterprise evolution
- Enable the pillars of a complete DevOps 360° operating model
- Adopt DevOps at scale and at relevance in a multi-speed context
- Implement proven DevOps practices that large incumbents banks follow
- Discover core DevOps capabilities that foster the enterprise evolution
- Set up DevOps CoEs, platform teams, and SRE teams

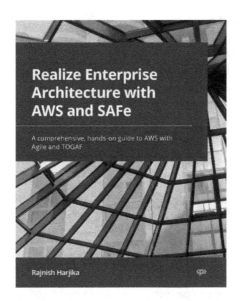

Realize Enterprise Architecture with AWS and SAFe

https://packt.link/9781801812078

Rajnish Harjika

ISBN: 978-1-80181-207-8

- Set up the core foundation of your enterprise architecture
- Discover how TOGAF relates to enterprise architecture
- Explore AWS's EA frameworks and find out which one is the best for you
- Use SAFe to maximize agility in your organization
- Find out how to use ArchiMate to model your architecture
- Establish proper EA practices in your organization
- Migrate to the cloud with AWS and SAFe

Packt is searching for authors like you

If you're interested in becoming an author for Packt, please visit `authors.packtpub.com` and apply today. We have worked with thousands of developers and tech professionals, just like you, to help them share their insight with the global tech community. You can make a general application, apply for a specific hot topic that we are recruiting an author for, or submit your own idea.

Share Your Thoughts

Now you've finished *Python Essentials for AWS Cloud Developers*, we'd love to hear your thoughts! Scan the QR code below to go straight to the Amazon review page for this book and share your feedback or leave a review on the site that you purchased it from.

https://packt.link/r/1804610062

Your review is important to us and the tech community and will help us make sure we're delivering excellent quality content.

Download a free PDF copy of this book

Thanks for purchasing this book!

Do you like to read on the go but are unable to carry your print books everywhere? Is your eBook purchase not compatible with the device of your choice?

Don't worry, now with every Packt book you get a DRM-free PDF version of that book at no cost.

Read anywhere, any place, on any device. Search, copy, and paste code from your favorite technical books directly into your application.

The perks don't stop there, you can get exclusive access to discounts, newsletters, and great free content in your inbox daily

Follow these simple steps to get the benefits:

1. Scan the QR code or visit the link below

https://packt.link/free-ebook/9781804610060

2. Submit your proof of purchase

3. That's it! We'll send your free PDF and other benefits to your email directly